DC POWER SUPPLIES

DC POWER SUPPLIES:

APPLICATION AND THEORY

ROBERT J. TRAISTER

RESTON PUBLISHING COMPANY, INC.
A Prentice-Hall Company
Reston, Virginia

Library of Congress Cataloging in Publication Data

Traister, Robert.
DC power supplies.

Includes index.
1. Electronic apparatus and supplies–Power supply–Direct current. I. Title.
TK7868.P6T72 621.31'042 78-31529
ISBN 0-8359-1275-2

A Prentice-Hall Company
Reston, Virginia 22090

10 9 8 7 6 5 4 3 2 1

Printed in the United States of America

To Robbie

CONTENTS

PREFACE

Modern power supply technology has done away with the bulky components of the past. Solid-state rectifiers have replaced the old diode vacuum tubes that consumed large amounts of filament power. Even transformers are smaller now, with the capability of handling larger amounts of current. Electrolytic capacitors have replaced the oil-filled paper variety and provide higher amounts of capacitance while occupying a smaller space. Today's dc power supplies no longer have the image of a massive compilation of steel, copper, and large hardware that must be mounted as close to the floor as possible to avoid disaster should its chassis give way under mechanical strain. They are lightweight, are often mounted on printed circuit boards, and form an integral part of electronic circuits.

This text will deal with the modern aspects of power supply circuitry using the components and technology that are relevant to today's design and building practices. Much of this circuitry is a direct result of experimental research made years ago with some of the obsolete components mentioned earlier but has been refined and improved to meet today's dc power supply requirements.

Robert J. Traister

DC POWER SUPPLIES

part one

BASIC THEORY AND APPLICATIONS

Chapters 1–8 of this text are an explanation of dc power supply circuitry presented in a logical manner, from the basic circuits on through more complex designs. Each section of the dc power supply is taken apart and described as to function and relative place in the overall circuit.

Although Chapters 1–8 deal mainly with theory and building practices, the material is presented in such a way as to make direct application of the principles learned possible at an early stage. This information should be well received by the individual who is relatively new to electronics, and should provide an excellent review to the more seasoned technician, as well as providing new information that will be necessary for future design and application. This information also makes an excellent source of practical reference material.

Chapters 9–11 use the information learned in the first eight. To skip lightly over the practical information in the first chapters will decrease the effectiveness of the various discussions provided in this text, and will also decrease the attained knowledge level, which the step-by-step discussions are designed to provide.

Study and retain all information provided. Answer the questions provided at the end of each chapter, which is best done by reading back through the chapter just completed. The answers to these questions may not be provided in the chapter, but the information to answer them by calculation and deduction is. Answer all study questions as fully and completely as possible. Any weak areas of study should dictate the rereading of the chapters involved until a good, working relationship with the formulas and information provided is obtained.

chapter one

POWER-SUPPLY CIRCUITRY

Most direct-current (dc) power supplies in use today are powered from an alternating-current (ac) source rated at 110 or 220 volts (V), with a frequency range of from 50 to 60 hertz (Hz). This ac source enters the power supply at the primary input to the power transformer, which will provide an ac voltage at its output of an amount that may range from a fraction of a volt to several thousand, depending on the type of transformer and its purpose in the supply circuit. The frequency of the secondary voltage will remain the same as that of the primary supply, but the voltage will be increased, decreased, or may even remain the same. The output voltage from the transformer secondary will determine the final dc output voltage from the supply after rectification and filtering have taken place.

RECTIFIER CIRCUITS

Figure 1-1 shows schematically the three basic rectifier circuits that are used individually and in combination in modern power-supply design. A *rectifier* is a device that will conduct current in only one direc-

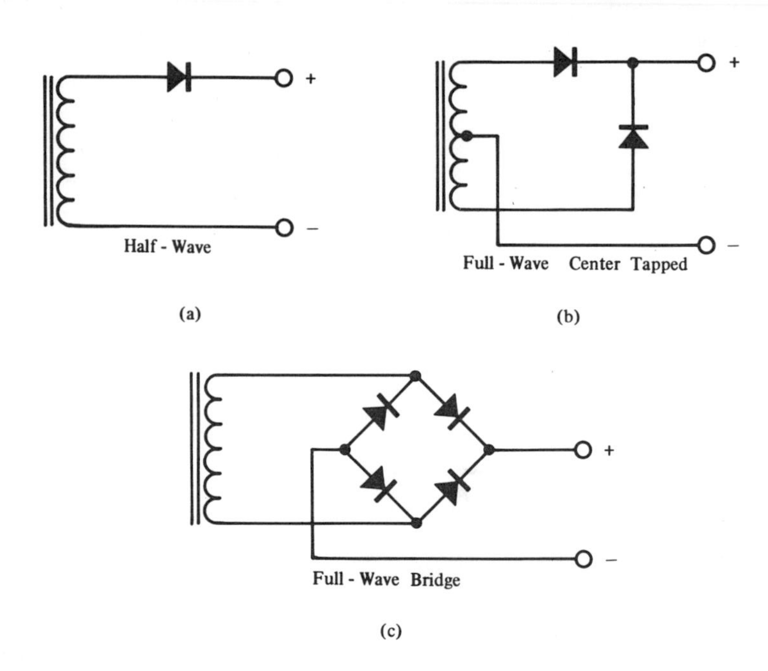

FIGURE 1-1. Rectifier configurations.

tion. Alternating current at a frequency of 60 Hz reverses its polarity at a rate of 60 times per second, but the rectifier will allow this current to pass in only *one* direction and, therefore, blocks one half of each cycle in the half-wave rectifier circuit show in Figure 1-1.

The ac waveform (Figure 1-2) is a sine wave that travels from zero to a peak positive value, returns again to zero, and then swings to a peak negative value before completing the cycle by returning to zero again. If the half-wave rectifier circuit is designed to pass only positive current values, the positive half of each sine wave will be allowed to

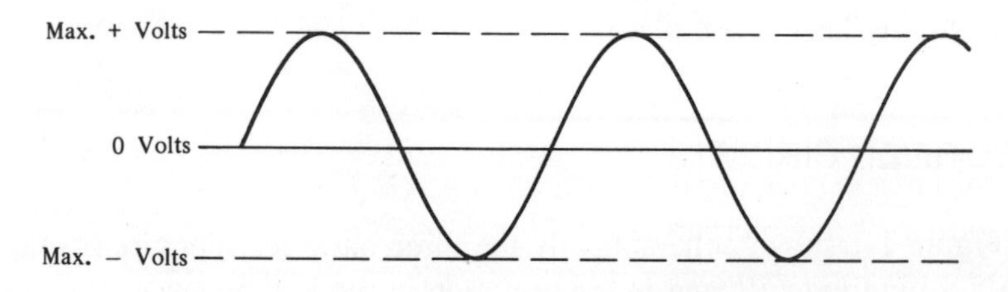

FIGURE 1-2. An ac sinewave pattern.

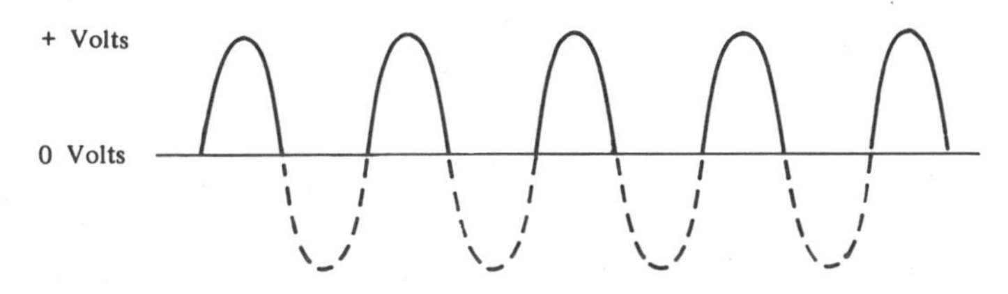

Blocked portion of former sine wave would appear here before rectification

FIGURE 1-3. Pulsating dc pattern.

continue through the circuit while the negative portion will be blocked entirely. Figure 1-3 shows the appearance of the wave after rectification has been accomplished. The current is now of a positive polarity only and is called *pulsating direct current* or *direct current,* because the voltage still rises from zero to a peak positive value and then back to zero again. The frequency of this pulsating direct current is still 60 Hz when using half-wave rectification.

Pulsating direct current is normally unsuitable for direct applications because of its unstable nature and variable degrees of voltage output. A filtering circuit is used at this point to smooth out the pulsations or ripple to produce a pure dc output. These filtering circuits will be discussed later.

Half-wave rectifier circuits are of limited use in practical application and are usually limited to simple, low-current devices where voltage stability is noncritical and power demand is relatively constant.

A widely utilized rectifier circuit is shown in Figure 1-1b, the full-wave, center-tapped configuration. This circuit is actually two half-wave rectifiers combined, and works by conducting both the negative and positive portions of the ac sine wave. A power transformer with a center-tapped secondary winding is a necessity for this type of rectifier circuit.

Since this circuit utilizes both the positive and negative portions of the sine wave in a combining or adding operation, the frequency of the pulsating dc output is twice that of the half-wave rectifier, or approximately 120 Hz. Higher frequencies require less filtering for a pure dc output when compared to lesser frequencies, which makes the full-wave center-tapped rectifier circuit more practical and economical than its half-wave counterpart when good regulation and voltage stability are a crucial requirement. The full-wave, center-tapped rectifier is the most universally used circuit design.

The full-wave bridge rectifier (Figure 1-1c) is the third circuit that can be considered to be a basic part of power-supply construction. As with the center-tapped rectifier, the full-wave bridge circuit operates on each half of the full ac cycle. Current flows through two rectifiers on the positive portion of the cycle and through the remaining two when the polarity swings to the negative side. Pulsating dc output is the same as obtained with the center-tapped circuit, and exhibits a frequency of about 120 Hz for an easier filtering requirement.

The proper transformer for use with the full-wave bridge rectifier circuit does not require a center-tapped secondary winding, which means that the entire output ac voltage of this secondary will be reflected in dc output. A full-wave bridge rectifier circuit will result in twice the voltage output in direct current than will the center-tapped circuit, because in the center-tapped arrangement only half the total secondary voltage is utilized with each portion of the ac cycle.

Each of the three basic rectifier circuits discussed has individual advantages and disadvantages. Voltage, current, and component requirements may dictate which should be used to best advantage in particular circuits with specified applications.

FILTERS

Pulsating direct current from the rectifier circuits of Figure 1-1 is not constant enough in amplitude for most practical applications. A hum that corresponds to the frequency of the pulsations will occur and may actually be heard when the power supply is used with audio electronic equipment. At this point in a power supply, filters consisting of capacitors and in some instances inductors or chokes are required between the rectifier and the load to smooth out the pulsations for a more consistent dc voltage. The design of filtering circuits will largely determine the amount of voltage regulation that will be exhibited by the supply, as well as the maximum amount of current that can be drawn without exceeding the ratings of the power-supply components.

Power-supply filters fall into two possible categories, choke input and capacitor input. Capacitor-input filters usually exhibit a relatively high output dc voltage in respect to the ac voltage of the transformer secondary; choke-input filters have less dc output. Capacitive-input filters are often used in applications that require a high peak voltage throughout the operation cycle, whereas choke inputs are used in applications that require constant voltage and current.

As more current is drawn from a dc power supply, the output voltage always drops. This is due in part to the resistance within the circuit wiring and in the windings of the power transformer, and because of the tendency of output voltage to soar during periods of very light current drain. When little or no current is drawn, dc output voltage will often be as high as one and a half times the secondary winding ac voltage. Proper design of filtering circuits will nearly eliminate this tendency during light loading, and will hold the dc output voltage to a more consistent level during conditions of varying current demands.

A filter circuit is normally designed to perform most efficiently while operating into a specified load. Circuits that draw current at a constant rate with little or no change do not require excellent *regulation,* a term used to describe the amount of voltage change under loading conditions. A circuit that demands heavy current during one portion of its operating cycle and light current during another portion will require better regulation to maintain the dc output voltage at a more constant level. Devices that draw large amounts of current for an instant, drop off to almost no current drain for a short period of time, and then repeat this cycle require the best regulation possible and a much more complex filtering circuit.

Capacitive-input Filters

Figure 1-4 shows schematics of capacitive-input filters commonly used in many practical power-supply circuits. All three types share the same characteristics, although the more complex circuits will provide better regulation. The single capacitor filter in Figure 1-4a is commonly used in many high-voltage power supplies where extremely stiff regulation is not required. Because of the low dc resistance of this circuit, a higher dc output voltage will result under light loading conditions. Figure 1-4b shows a single-section or "PI" filter circuit, which is similar to the capacitor filter mentioned previously, but with the addition of a choke or inductor and another capacitor. Due to the added dc resistance through the choke windings, dc output voltage will be slightly lower; but regulation is much better, and output voltage will remain more constant under varying load conditions. Figure 1-4c depicts a double-section filter, which adds another choke and capacitor to a single-section filter to provide even stiffer regulation and ripple re-

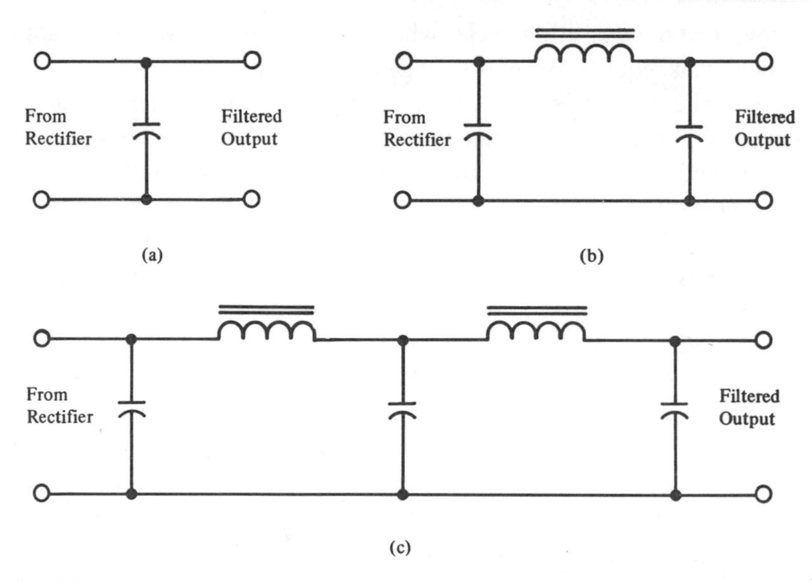

FIGURE 1-4. Capacitive input filter circuits: (a) single capacitor; (b) pi-section; (c) double pi-section.

duction. *Ripple* is a term used to describe the ac component that is present in the dc output of a power supply.

Filter construction is usually the most critical and crucial portion of correct power-supply design. Filter characteristics will usually determine which specific applications the power supply is suitable for.

Choke-input Filters

Choke-input filters have limited use in power-supply circuits of modern design. Filters of this type worked best when tube-type thermionic rectifiers were in great use.

A choke-input filter will exhibit much the same characteristics as its capacitor-input counterpart unless the input choke has a specified minimum value of inductance, called *critical value*. This critical value of inductance, which is measured in henrys, is determined by the operating voltage and current that the power supply must provide. When this critical value is met, dc output voltage will be roughly equivalent to the *average* ac voltage produced across the transformer secondary. In practical application, the dc voltage will be about 10 percent less than the average ac voltage owing to circuit resistance.

Chokes are carefully chosen for critical value and voltage–current ratings, which are determined by the demands to be made on the power supply. The choke must have the correct inductance to maintain its critical value, the current-handling capacity to prevent overheating when power is drawn, and the voltage rating to prevent arcing between the positive and negative portions of the power supply. Chokes for power-supply applications are usually of the iron-core variety and resemble power transformers in physical construction. They are normally mounted directly to the power-supply chassis, which is usually the dc ground. The amount of voltage that can be placed across the choke winding before an arc will occur between it and the choke casing determines the voltage rating for this device. Voltage ratings for many electronic components usually contain a significant safety factor, and can sometimes be exceeded without a major breakdown, but this practice is not dependable nor safe and all manufacturers' maximum ratings should be adhered to at all times.

Figure 1-5 illustrates a typical choke-input filter. The chokes are placed in the positive lead of the power supply shown at Figure 1-5a; however, in many applications, a choke placed in the negative lead will perform as well, and the voltage rating of the choke may be much lower because its case and voltage windings are operated at or near ground potential (Figure 1-5b).

A *swinging choke* is a power-supply choke that will maintain critical value of inductance over a large range of current drawn from the supply. It exhibits a variable inductance that is determined by the current drawn from the power supply. Swinging chokes are not usually suitable for critical uses that require high regulation and stability, but they are inexpensive when compared to standard chokes and still have limited usage.

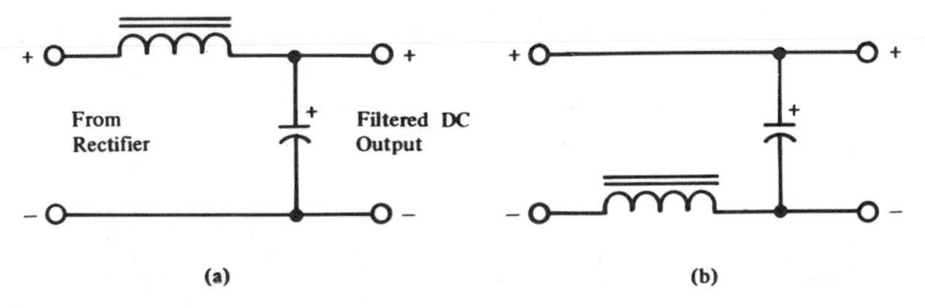

FIGURE 1-5. Choke input filter circuits: (a) single section filter; (b) negative line filter.

BLEEDER RESISTORS

The final basic component in the simple power-supply circuit is the bleeder resistor, so called because it bleeds or drains the stored power in the capacitor or capacitors that compose the filter circuit. Without the bleeder resistor, capacitors, which have the ability to store current for long periods of time, would remain charged and pose potentially lethal hazards even after the power supply was disconnected from its ac power source. Another function of the bleeder resistor, which is connected across the positive and negative outputs from the power supply, is to improve voltage regulation by providing a definite load. The resistance of the bleeder resistor can be considered the minimum load resistance. During absolute no-load conditions, it has been learned that dc output voltage tends to rise sharply; but with this minimum load connected in the form of a bleeder resistor, the power supply always sees an amount of load resistance and sharp voltage rises can be lessened.

There are specific formulas for calculating the correct resistance in ohms for specific dc voltage outputs, as well as the power rating of the resistor in watts. Typically, power ratings are at least doubled in choosing the resistor to allow a sizable safety factor. If the resistor should become overheated and fail to perform its function in the circuit, dangerous voltages could remain in the circuit after the supply is switched off.

Figure 1-6 shows the schematic position of the bleeder resistor in the power-supply circuitry. As an added safety factor, many high-voltage power supplies will have resistors of very high resistance paralleled with the proper bleeder resistor. Should the main resistor malfunction for any reason, this auxiliary resistor would slowly drain any remaining current in the circuit.

Bleeder resistors will give off large amounts of heat in medium- to

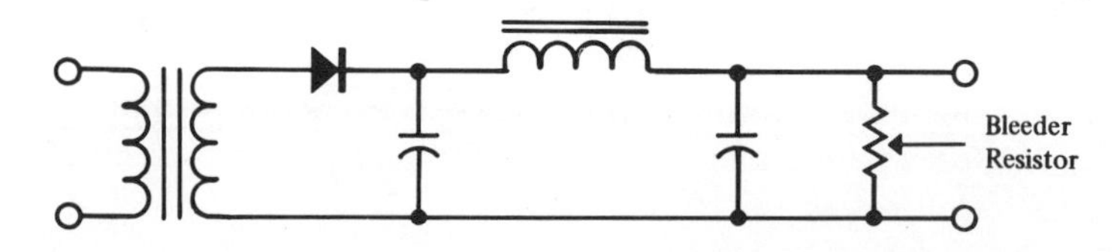

FIGURE 1-6. Simple power supply circuit showing proper circuit location of the bleeder resistor.

high-voltage power supplies and must be mounted away from any heat-sensitive components in the circuit. Some high-voltage power supplies even have a small fan or blower that effectively removes radiated heat from these resistors.

The safety factor is a much stressed point when discussing bleeder resistors. Moderate voltages of less than 300 can cause severe electrical shock and death in some cases, but there is no such thing as a small shock from a power supply rated at voltages of 2,000, 3,000, 5,000, and more. To come in contact with voltages of this magnitude can easily mean death or crippling. Bleeder resistors of proper ratings coupled with intelligent operations regarding power-supply circuitry prevent severe electrical mishaps from occurring.

AC LINE CONSIDERATIONS

Now that the basic circuitry and components that comprise the dc power supply have been discussed, the source of this finished product, the ac power source, can be studied as to its usual properties and its effect on the dc output and operation of each component in the power supply.

Most power supplies operate from a voltage of 110 or 220, which is often referred to as *house current*. Different areas may refer to house current at different ratings such as 115–230 V, or even 120–240 V, because the value of the line voltage may vary somewhat. Many transformers even have different primary taps to allow more precise control over the output voltage; however, in most applications, this difference will have little effect. Most power companies try to maintain a voltage of 115–230 as the standard voltage value.

The *line frequency,* or the rate at which the current polarity reverses, is maintained at 60 Hz in the United States, but may vary slightly from location to location. Line frequency is an all-important factor in power transformer design. A transformer that delivers a specified voltage at an input frequency of 400 Hz, for example, will appear as a short circuit at a 60-Hz rate and cannot possibly operate. Most transformers for use in practical application today are designed to operate from 60-Hz current. Transformers for other frequencies are usually specially ordered. Four-hundred-hertz equipment was popular on aircraft and still has some application, because components designed for this frequency of operation are smaller and weigh less than their 60-Hz counterparts.

The higher value of voltage, 230, is often used with power supplies that draw high amounts of current from the primary line. Half as much current is required at 230 V to supply the same amount of power output as would be obtained from 115 V. The lower current is more easily conducted through the power wiring without incurring high ohmic or resistance losses, and better regulation is the end result as primary voltage is maintained at a steadier rate. Just as current demands from the dc power supply meet resistance within the circuit components, ac current meets with resistance and loss within the house wiring, incurring a loss of voltage. An indication of excessive voltage drop is seen in the dimming of lights on the same circuit as the power supply.

Another characteristic of the ac line is that of *phase,* which is another description of time. House current is always single phase in nature. Large factories and business complexes often use a three-phase system for its increased power-handling capabilities. For this practical study, all circuits discussed will derive their primary voltage from single-phase electrical systems.

In certain areas, problems may be encountered with the line voltage dropping to a low rate. A change of a few volts is not usually noticeable in other than precise applications, but when a drop of 10–20 V occurs, transformers and chokes may become overheated, depending on the type of load. Some transformers offer a primary winding that is tapped at several different points. These taps are used when different voltages must be used for the primary ac source. For example, a transformer that delivers 500 V at the output of its secondary winding may contain taps on the primary winding that will deliver the rated voltage output when connected to a 100-, 105-, 110-, 115-, 120-, or 125-V source. When the value of the line voltage is determined by measuring, the primary tap that falls closest to the actual line voltage is used to obtain the 500-V output. If actual line voltage is measured at 103 V, the 105-V tap would be used for the ac input, because it falls closest to the actual line voltage value.

In many instances, voltage drop will occur suddenly and remain in a low state for some time before returning to a normal value. This drop may occur in the evening when heavy-current devices such as stoves are turned on for cooking, or during periods of very hot weather when air conditioners, more heavy-current devices, are in use. A tapped primary winding on a power transformer is impractical for use in a situation such as this because the low-voltage condition is not continuous. The voltage connection points to the transformer primary would have to be changed as the line voltage rises and falls.

Continually varying line voltage conditions may often be overcome through the use of a constant-voltage transformer. These devices are relatively expensive, but perform an excellent job of maintaining line voltage at a steady value. Constant-voltage transformers are available in many sizes and power output ratings, contain no moving parts, and require no adjustments after initial setup has been completed.

Another expensive but useful device for maintaining proper line voltage is the variable autotransformer, sometimes referred to as a variac. When inserted between the ac line and the power supply or other equipment to be powered, the variac allows a continuous manual adjustment of ac line voltage. A variable control is used to increase or decrease the line voltage and may be done in seconds. Although less convenient than a constant-voltage transformer, the variac, when coupled with a line voltage meter, can prove a significant advantage in dc power-supply applications under conditions of abnormal line voltage values.

BASIC COMPLETED DC POWER SUPPLY

By combining the various portions of the dc power supply, the completed unit is arrived at. Figure 1-7 is a schematic of the sample circuit; it consists of a full-wave center-tapped rectifier, a capacitive-input filter, and bleeder resistor. This particular circuit may be designed to deliver any amount of voltage and current when the proper component values are chosen.

By referring to Figure 1-7, each step in the process of supplying dc current to the load may be followed in sequence. Alternating line current is fed to the primary winding of the power transformer, where it is

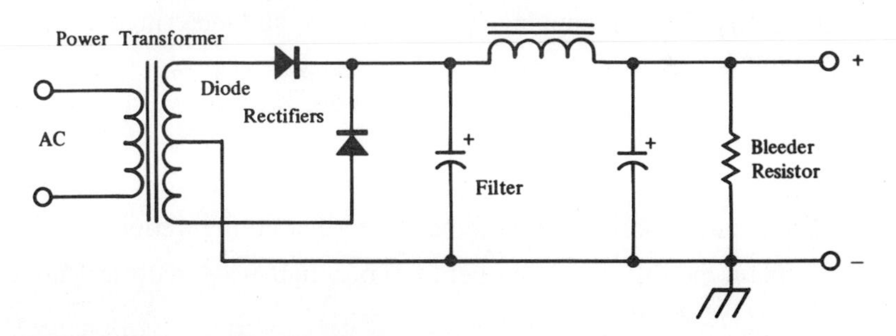

FIGURE 1-7. Completed power supply using full-wave, centertrap rectifier, capacitive input, pi-section filter and bleeder resistor.

changed to output at the secondary winding. The full-wave center-tapped rectifier assembly changes the alternating current from the secondary winding into pulsating direct current. The capacitive-input filter smooths out the pulsating direct current for a pure dc current output to the bleeder resistor and the load.

Although different components and circuitry may be used, virtually every dc power supply follows this line of progress in delivering pure dc current to the load. The steps are (1) voltage transformation, (2) rectification, and (3) filtering. More advanced power-supplies may add electronic regulation circuitry, but every supply must include these three steps for a nonpulsating dc output.

When studying more advanced power-supply circuitry, remember that, no matter how complicated the circuit may appear, the three basic power-supply portions are always within the overall circuit, and can be dealt with on an individual basis for a better understanding. Much of the electronic circuitry follows this same pattern. The most complex pieces of equipment can usually be broken down into a few simple circuits that are duplicated in many ways throughout the entire circuit.

SUMMARY

This chapter has dealt exclusively with *basic* dc power-supply design through discussions on segments of the overall supply circuitry. Specific components as well as their values and ratings will be discussed in later chapters. The information provided is essential for any further study in power-supply circuitry. Remember the three basic power-supply portions and prepare to refer back to them constantly in all studies of dc power-supply circuits and components. All power supplies studied in this text will start with these basic circuit segments; then protective devices, regulatory components, and metering circuitry will be added.

QUESTIONS

1. Name and describe the three basic rectifier configurations.
2. Explain the difference in operation of a half-wave rectifier and a full-wave rectifier.
3. What kind of waveform is supplied by alternating line current?

4. What is pulsating direct current?
5. How does pulsating direct current differ from alternating current?
6. Name the two basic types of filter circuits.
7. What are the characteristics of the two basic filter circuits regarding dc output voltage?
8. Under what conditions will a filter circuit perform most efficiently?
9. How is the inductance of a filter choke derived?
10. What is critical value?
11. Describe the differences between standard chokes and swinging chokes.
12. What purposes do bleeder resistors serve in a dc power supply?
13. Why can a power supply with no primary voltage being supplied be a safety hazard when the bleeder resistor is omitted?
14. What are the characteristics of standard line voltage with regard to average voltages, frequency, and phase?
15. What advantage is there in using higher line voltages when supplying power to high-current devices as compared to the lower voltages encountered in house current?
16. List three ways voltage-fluctuation problems may be corrected.
17. What are the three basic dc power-supply portions?
18. What is a component safety factor?
19. What is the primary winding of a power transformer?
20. What are the three steps to dc power-supply output?

chapter two

POWER-SUPPLY COMPONENTS AND RATINGS

Now that a basic understanding of power-supply circuitry has been gained, the individual components will be examined as to construction, function(s), ratings, and uses in each segment of the overall circuit.

TRANSFORMERS

Transformers used in conventional dc power supplies are usually of the iron-core variety. The primary and secondary windings of the transformer are wound on a single iron core, which is normally composed of laminated plate sections stacked one atop the other to form the core thickness. Figure 2-1 shows an example of a typical iron-core-transformer arrangement. The physical appearance of different cores may vary, but the basic principle of construction is the same.

Transformers perform their functions by transferring electrical energy from one circuit to another. The primary winding and the secondary winding or windings share a mutual inductance, which allows this transfer to occur without the necessity of a direct connection. A transformer can operate only with alternating current, which has a con-

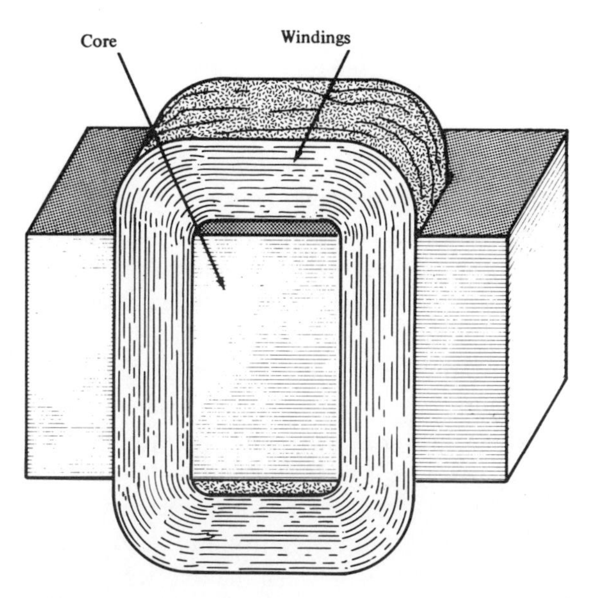

FIGURE 2-1. Pictorial diagram of power transformer.

stantly changing magnetic field due to the change in polarity, which identifies an ac waveform. If direct current is applied to the primary of a transformer, a voltage will be present across the secondary winding only at the instant that it is applied or removed, since this is the only time that a change in the magnetic field is present with direct current. If direct current is fed to a transformer primary for more than an instant, excessive heating will occur within the windings, causing rapid destruction of this electrical device.

It is possible but not practical at ac frequencies to build a transformer without an iron core. However, the physical size of this device would be enormous in most cases owing to the loss in inductance. When a transformer is composed of windings mounted on a magnetic material such as iron, the overall value of inductance for the number of turns is multiplied. The iron core does present certain problems because of small currents that flow within the iron core itself. These are referred to as *eddy currents,* and they set a limit on the amount of current that can be supplied by the transformer. Other factors, such as wire size of the primary and secondary windings, also set a limit on the current that can be drawn, and must be considered when choosing a transformer for any power-supply circuit.

The main function of a transformer used in dc power-supply circuits is to change one voltage to another. This change may increase or decrease the applied voltage or keep it the same, while isolating it from its source owing to the fact that there is no direct connection of the primary and secondary windings.

Construction of power-supply transformers is tailored along design lines that usually keep the magnetic path around the core as short as possible. The shorter the magnetic path, the fewer the number of turns needed in the primary and secondary windings. Fewer turns in these windings reduce by a considerable amount the resistance or ohmic losses, and the entire unit will function more efficiently, with less power being dissipated or thrown off by the core material.

Referring to Figure 2-2, two typical transformer core arrangements are shown. Figure 2-2a shows the shell design; both the primary and secondary transformer windings are placed on the inner portion of the core or leg. The true *core* type of transformer arranges the primary and secondary windings on separate legs (Figure 2-2b). The latter design minimizes capacitance effects between the two windings by providing a greater amount of physical spacing between them. This spacing also provides a much higher insulation factor and is ideal when one of the windings must operate at a very high voltage. The chance of voltage arcing between the primary and secondary winding is decreased. Other than the differences noted, both the shell design and the core design of transformer construction operate similarly.

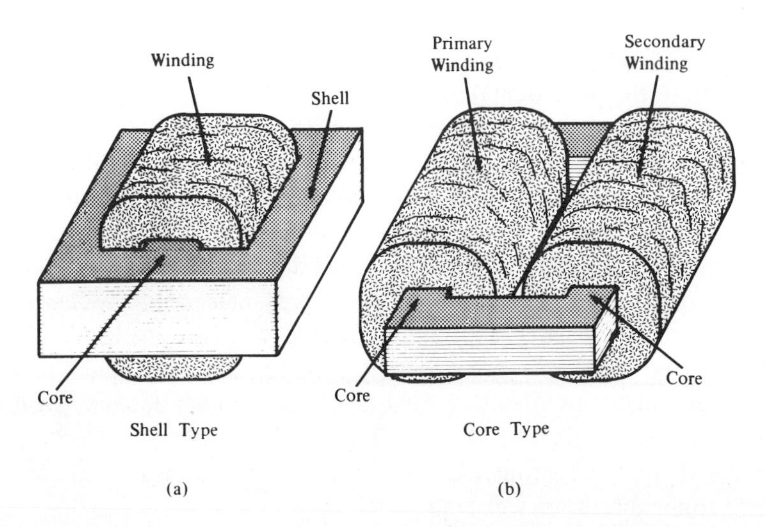

FIGURE 2-2. Pictorial diagram of two transformer core arrangements.

Construction of transformer cores used in most power-supply transformers is accomplished by using small sheets or sections of silicon steel. These sheets or laminations are painted with an insulating material and then are stacked together, one atop the other, to form the bulk of the core. Insulated sections decrease the flow of eddy currents, which will flow in a solid, uninsulated core and cause increased heating effects and losses.

Again, the number of turns required in the primary and secondary transformer windings is directly related to the size, shape, and construction type of the core, which includes the type of core material or materials used in construction. Another determining factor is the frequency of the applied alternating current. Calculations on turn windings are figured in turns per volt, and are obtained by computing the cross-sectional area of the core. A longer magnetic path or smaller cross-sectional area of the core will require a larger number of turns. A smaller path or larger cross-sectional area will require fewer turns. At a line frequency of the standard 60-Hz, many power-supply transformers using standard core materials and construction will require about 8–10 turns per volt.

The *volt–ampere rating* of a power transformer, or the amount of power that it will deliver while still maintaining a safety factor, is dependent on several factors, which include size of core, conductor size of primary and secondary windings, and the type of filter and rectifier designs used. A capacitor-input filter delivers high peak values of voltage and current, which causes more of a heating effect in the secondary transformer winding, whereas the average values of voltage and current obtained with a choke-input filter will bring about cooler operation in comparable circuits. When using a capacitive-input filter, a larger transformer rating is often necessary, because the power delivered to the load may be considerably less that the power drain from the transformer.

Many power-supply transformers will have one primary winding and several secondary windings that provide different amounts of voltage and have different current ratings. The transformer may have one relatively high voltage winding, a medium-voltage winding, and still another with a relatively low voltage output. When using a full-wave center-tapped rectifier configuration, the transformer secondary must have a secondary winding that is tapped at its center. Figure 2-3 gives a schematic of this type of transformer. Technically, the secondary is composed of but one winding, which is tapped at the center with a lug or conductor that can be utilized in a full-wave center-tapped rectifier arrangement. The amount of voltage from the center tap to either the upper or lower end of the transformer is identical, and the dc output voltage will be decided by this ac voltage across one half of the entire secondary winding. This same transformer could be used in a full-wave bridge rectifier circuit or even a half-wave circuit by not using the center-tap connection. When using this type of circuit configuration, dc output voltage would be calculated from the ac voltage across the entire secondary winding of the transformer.

Transformers are normally rated by voltage output, current, and, in some instances, insulation to ground. This latter rating is not in great use today; it was intended to set maximum operating voltage limits for transformers that supplied filament voltage for the tube-type rectifiers used in the past. Some transformers may be rated in volt–amperes or

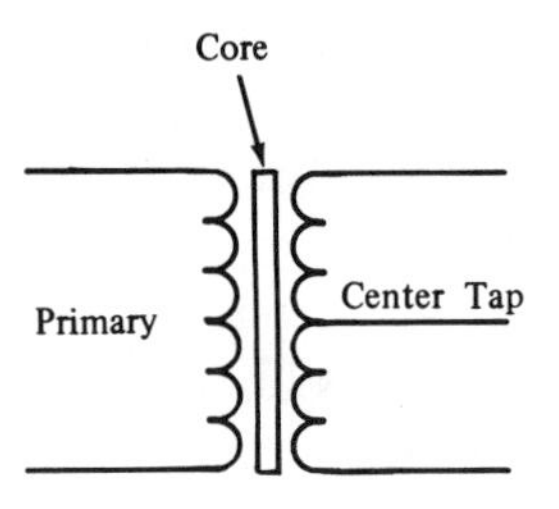

FIGURE 2-3. Schematic representation of power transformer with a center tapped secondary winding.

watts, with individual ratings compiled for both the primary and secondary windings. This is especially true of transformers of the multipurpose variety that can be used in both full-wave bridge and center-tap configurations. Some ratings on transformers and other electronic components may be labeled CCS or ICAS. The first term means *continuous commercial service,* which implies a more rugged use of the component than the latter term, which stands for *intermittent commercial and amateur service.* CCS indicates that the ratings supplied are applicable for round-the-clock continuous use; ICAS ratings are for short periods of use with equally long periods of inoperation. The latter ratings are normally higher for a specific transformer because intermittent use with periods of inoperation usually allows for better cooling of the transformer core and windings.

SOLID-STATE RECTIFIERS

A rectifier is a device that will conduct current of a specific polarity, either positive or negative, in one direction but not in the other. Solid-state rectifiers are made from germanium or silicon crystal materials, with silicon being used almost universally for power-supply applications.

Solid-state rectifiers are rated in many ways, but there are three values of interest in dc power-supply applications. The *peak reverse voltage* (PRV) is the voltage that the diode must be able to withstand when it is not conducting current. In application, this current will vary with the load and can be almost three times the ac voltage delivered by the transformer secondary, depending on the type of rectifier circuitry used. The second important rating in rectifier selection is abbreviated I_{rep}. This is the *peak current rating* of the rectifier, and is the maximum rating of current that can be passed through the device. The third rating is the *surge current,* sometimes listed as the I_{s} factor, which may be several hundred times the I_{rep} rating. I_{s} ratings set the maximum surge that the device can safely handle. This high current occurs for a small fraction of a second when voltage is initially applied to a dc power-supply circuit. An average current rating is also applied to solid-state rectifiers and sets the proper continuous operating limits of the device.

Solid-state rectifiers are very advantageous when used in the construction of dc power supplies. They require no external power, are very efficient, and are very compact when compared to the vacuum-tube rectifier of a few years ago, which required filament voltage, larger amounts of mounting space, and sometimes external cooling devices such as a fan or blower to remove heat.

Silicon rectifiers are available in a wide range of current and voltage ratings, with devices of less than 600-PRV rating capable of handling up to 400 amperes (A) and more. Higher-voltage rectifiers of around 1,000-PRV rating may have current ratings on the order of 1–2 A. Owing to the small physical size of solid-state rectifiers, it is possible to stack the units to provide current ratings and PRV ratings in multiples of the discrete component ratings. Commercially, stacked units are available with ratings on the scale of 15,000 peak reverse voltage (PRV) with current-handling capabilities of 1 A.

Other commercial packages come in the forms of one-piece full-wave center-tapped and full-wave bridge rectifier units, which have the leads protruding from the case and marked for appropriate connections to the power-supply transformer and filter circuits.

RECTIFIER PROTECTION

A silicon rectifier allows current to flow for only one half of the ac cycle, so when it does conduct, it may pass at least twice the average direct current. With a capacitive-input filter, it may pass peak current of 10 or more times the average direct current. This peak current, I_{rep}, is also known as the peak repetitive current, and, as explained earlier, is much less than the I_s or surge current rating, which applies to the surge current that flows through the rectifier when the circuit is initially switched on. One other current rating for solid-state rectifiers is the I_0 rating, or the average dc current rating. This sets the limit on the amount of current the rectifier may safely handle for continuous periods of operating time.

To sum up these current ratings, I_0 sets the limits on average current through the rectifier, I_{rep} limits the instantaneous peaks of current that may occur at intervals throughout the operating cycle, and I_s limits the instantaneous surge of current during initial circuit activation. The duration of time involved in the surge-current rating is for one ac cycle at 60 Hz, or about sixteen thousandths of a second.

Comparing these current ratings, a ratio of approximately 4:1 is obtained for I_{rep} and I_0. Thus if the I_0 rating is 1 A, the I_{rep} rating will be about 4 A. The ratio of I_s to I_0 is about 12:1, which would be approximately 12 A in a rectifier with an I_0 rating of 1 A.

Thermal protection of diodes is necessary in certain applications owing to the small physical size of these components. Fortunately, solid-state rectifiers have a low internal resistance, and heat problems do not occur in units rated for less than 2 A if operated within their ratings; but rectifiers rated for 2 or more amperes usually must depend on external heat sinks in order for maximum ratings to apply.

A heat sink is an external, usually metal, device that attaches directly to the solid-state component. It is a solid piece of metal that often has metal fins or vanes protruding from each side in order to release more heat into the air. The heat sink effectively increases the physical size of the solid-state rectifier, thus allowing a more efficient method of channeling heat away from the delicate crystal interior. Heat sinks are usually insulated from other portions of the circuit because, in most instances, the full dc voltage will be present on their exterior surfaces. If this is not practical, the rectifier will be insulated electrically from the heat sink but not thermally. A small piece of plastic or mica sheet will be placed between the rectifier and the heat sink for the electrical insulation. To ensure proper conductance of heat from the rectifier to the sink, silicone grease is sometimes applied to both units at the insulating material where they meet. Silicone grease is an excellent conductor of heat, and at the same time maintains a good electrical insulation property. The heat is conducted to the heat sink while the current is limited to the surface of the rectifier only. The heat sink may now be grounded to the common ground or chassis without danger of an electrical short circuit.

In extremely high current applications, small fans or blowers may be used to remove heat from solid-state rectifiers, but these applications are usually limited to industrial uses. Large heat sinks are attached in the way just described, with the forced air from the fan or blower directed across the cooling vanes and the solid-state device for more efficient removal of heat into the surrounding air.

As added protection for small current-handling rectifiers of less than 2 A that must be operated at maximum ratings for long periods of time, small clip-on metal tab heat sinks may be utilized. Some of these are meant to be bolted to the chassis of the power supply for additional transfer of heat. Normal insulating steps are taken with these attachments as with the others mentioned for optimum electrical blockage and heat conductance.

Transient Protection

A frequent cause of rectifier failure in dc power-supply circuits is transient voltage or spikes in the ac power line. These are quick surges of voltage that cause the transformer output voltage to be much higher than that normally applied to the rectifier. Spikes can be caused by high-current devices such as motors operating on the same ac line being switched on and off or even by distant lightning strokes; whatever their cause, transient voltage problems can cause permanent damage to solid-state rectifiers.

Suppressing line transients is easily accomplished. Figure 2-4 shows two ways. In Figure 2-4a, two capacitors are used, one in the primary input to the secondary, the other across the secondary. At 60 Hz, the primary winding capacitor looks like a very high resistance and has

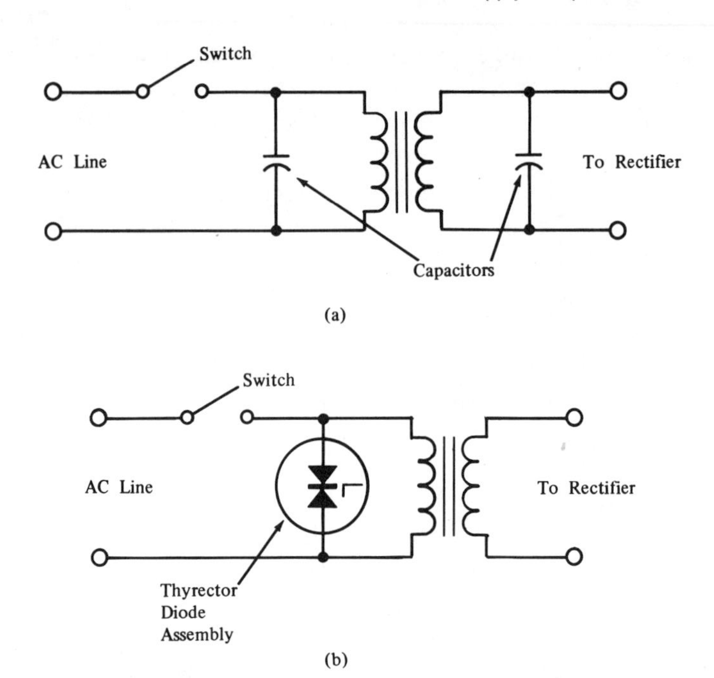

FIGURE 2-4. Spike and transient protection circuits for power supplies.

almost no effect on operation of the circuit; but when a spike occurs, its high-frequency component sees the capacitor as a short circuit and cancels itself before entering the transformer primary. The capacitor in the transformer secondary is an additional protection device, working in much the same manner as the primary capacitor.

Figure 2-4b uses a pair of selenium suppressor diodes, which do not conduct unless the peak line voltage becomes unusually high in value. When transients appear, the diodes conduct and, again, short out the spikes. This type of suppression device is often called by its trade name, Thyrector.

The two types of transient voltage protection shown do an adequate job of eliminating spikes from the dc power circuits, but neither is 100 percent effective in all cases. If persistent problems recur, rectifiers with double the PIV rating normally called for should eliminate further problems if used with these forms of spike protection.

Surge Protection

As was mentioned earlier, each time the power supply is turned on, the rectifiers see what appears to be a short circuit for the instant when the filter capacitor is charging. During this fraction of a second, large amounts of current are drawn through the rectifiers. In many cases and applications, this current may exceed the I_s rating of even the finest rectifiers and cause eventual failure. Surge protection is usually necessary to protect the diodes until the capacitor has been charged and normal, continuous operation characteristics of the power supply are in effect.

One method of reducing the surge potential when the power supply is activated is to install resistors of a relatively low ohmic value in each lead coming from the secondary of the power transformer, as depicted in Figure 2-5. This circuit configuration works well in power supplies that are required to deliver a continuous value of voltage and current; but when varying amounts of current are drawn throughout the operation of the supply, the resistors will cause poor voltage regulation. Surge protection devices in the primary circuit of the power supply will eliminate this drawback and provide adequate protection from current surges.

Figure 2-6 depicts a typical primary surge protection circuit. The surge-limiting resistor is located in the ac line to the power trans-

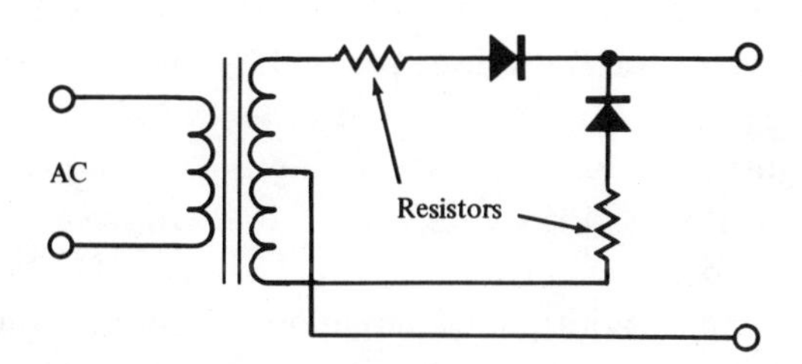

FIGURE 2-5. Surge protection using series resistors.

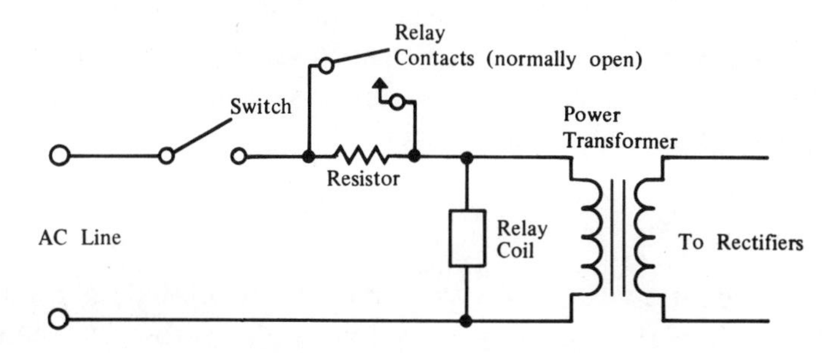

FIGURE 2-6. Surge protection using series resistor in the primary line and line-activated relay.

former and is electrically removed from the circuit a fraction of a second after the supply is activated and the filter capacitor is fully charged. The line resistor causes the input ac voltage to drop when the circuit is initially activated until the filter capacitor is nearly charged. When this capacitor approaches a full charge, the surge current drops and allows the relay in the primary circuit to activate. When this happens, the relay contacts short out the surge resistor, and for the rest of the operating cycle, the entire circuit functions as if the resistor were not there at all. This entire process takes place in a fraction of a second, and the delay of the relay in activating cannot usually be detected. A relay with contacts rated to handle the ac current drawn by the supply must be chosen, because the full current demand is conducted across these contacts. The value of the surge resistor will vary depending on the type of supply and the primary voltage, but resistance will usually be between 10 and 75 ohms (Ω) with a power rating of 5–50 watts (W).

RECTIFIER STACKS

As was previously discussed, solid-state rectifiers have specific limits in voltage and current ratings per unit. Owing to cost and physical space factors, it is often convenient to stack rectifiers in series or parallel configurations for increased ratings. Two diodes in parallel exhibit the same PIV or PRV ratings, but forward current may be almost doubled. Rectifiers in series still maintain the same current ratings as would apply with a single component, but the voltage ratings may be almost doubled. Owing to differences in internal resistance, special circuitry should be provided to make certain that one rectifier does not pass the majority of the forward current or be subjected to the majority of the peak inverse voltage. When internal resistances differ, either of the two conditions could exist, resulting in one or more rectifiers handling the majority of the work and eventually failing in conditions beyond normal operating limits.

Diodes in Series

When two or more solid-state rectifiers are placed in series, a capacitor and resistor should be placed in parallel across each component in the stack to equalize the PIV drops and as further protection against voltage spikes. When two resistors are combined in parallel, the resulting total resistance is less than the value of either resistor. By thinking of the internal resistance of the rectifier as a resistor, the added external resistor, which is of a much higher resistance value, will equalize the internal resistance to a value almost equal to the next diode–resistor combination in the string or stack (see Figure 2-7a). The value of the external resistor is not critical, but a means of calculating the resistance can be had by multiplying the PIV rating of the diode by 500, with the resulting figure being the ohmic value of the external resistor. A rectifier with a PIV rating of 1,000 would require a shunting resistor of approximately 500,000 Ω or 500 kilohms (kΩ). This calculation holds true whether there are 2 or even 20 diodes in the stack.

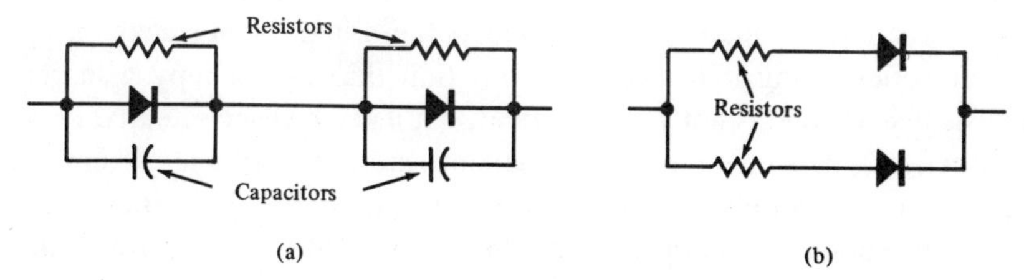

FIGURE 2-7. Protective circuits for diodes in series and in parallel.

The capacitors aid the rectifiers in the string to conduct and to block current flow at the same time and should be of a value of about 0.01 microfarad (μF). Capacitors used for this purpose should be of the noninductive ceramic-disc type, with a voltage rating equivalent to the voltage dropped at each rectifier. If two diodes are used in a power supply with a 500-V dc output, the protection capacitors should be rated higher than half of this amount, since the total output is divided by two, the number of diodes in the string. Doubling the ratings of these capacitors will assure adequate protection from voltage breakdown. Capacitors should be of the same value and manufacture, with 10 percent tolerances usually termed as adequate. All diodes in the string should also be of the same type and manufacture to assure proper equalization.

Diodes in Parallel

For greater current-handling capabilities, diodes may be placed in parallel provided adequate equalization procedures are taken. Figure 2-7b shows how two equal-value resistors are placed in series with the rectifiers and in parallel with each other. If the resistors are omitted from the circuit, one diode may draw most of the current and malfunction after a brief period of time. The resistors equalize the internal resistance of the stack and are of a very low ohmic value. Resistor values are chosen to deliver a voltage drop of about 1 V at peak current demand from the supply using Ohm's law of $E = IR$, where E is the voltage drop of 1 V and I is the peak current. R will then be the required value in ohms of the equalization resistors.

Other Combinations

Figure 2-8 shows a combined series–parallel configuration using solid-state rectifiers and equalization–protection components. Here the stack contains paralleled rectifiers and equalization resistors for greater current-handling capability and connects these individual stacks into a larger series string. This is a combination that can occupy a larger physical space and is not very practical, but it can be used effectively in a situation where the correct components are not available. Examine this schematic closely. It can be seen that both forms of rectifier combination, series and parallel, have been used with proper protection circuitry.

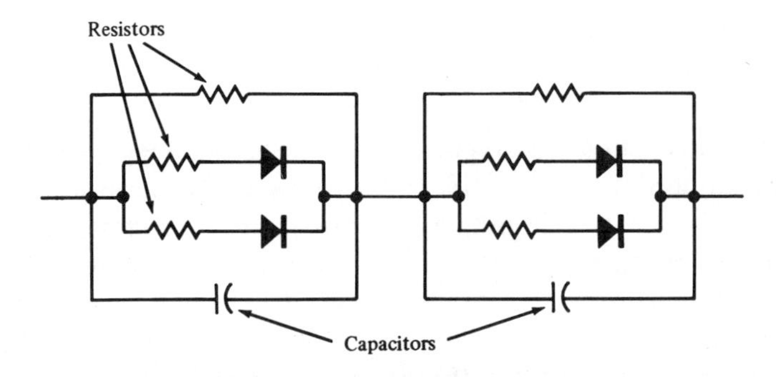

FIGURE 2-8. Series-parallel combinations and associated circuitry.

FILTER COMPONENTS

Pulsating direct current is smoothed out by the filter circuitry to obtain a pure dc waveform at the power-supply output. Filters are made up of capacitors and sometimes chokes. Proper filter construction will likely determine the output voltage and current that the supply will deliver and will also have a large effect on the regulation of the delivered product.

Capacitors

Capacitors used in dc power-supply applications have the ability to store electrical current and help to smooth out the pulsating dc wave by adding voltage to the circuit when the wave starts to decline. Electrolytic capacitors are almost exclusively used in all but the very high voltage power supplies because of their high-capacitance- and small-physical-size properties. Electrolytic capacitors are polarity sensitive devices and must be connected properly in the circuit to prevent component damage. Many electrolytic capacitors contain a liquid insulation chemical called a *dielectric,* which can overheat when correct polarity is not observed. This overheating causes the interior of the capacitor to expand and the entire component can explode. Most electrolytics have a pop-out pressure release port, but these do not always function properly; correct polarity connections are a must for safety as well as for proper circuit operation.

One disadvantage of electrolytic capacitors is a relatively low voltage rating when compared to the older type of paper capacitors. This may be overcome by forming a series string. When two capacitors are connected in series, like rectifiers, their voltage rating as a unit is twice that of a single component, assuming that both capacitors have equal ratings. However, when capacitors are placed in series, their total capacitance is less than that of a single component. If two capacitors rated at 400 V and 20 μF each are placed in a series string, the total voltage that can be applied to the combined units would be 800, while the total capacitance exhibited by this string would be a little less than 10 μF, Figure 2-9 shows a schematic of capacitors in series, along with formulas for calculating voltage rating and effective capacitance.

To increase the effective capacitance of capacitors, parallel circuit strings may be considered. Figure 2-10 shows that the capacitance of all the units utilized in the string is added, but the voltage rating is no greater than the lowest rating of any one capacitor in the string.

These principles of capacitance apply to all types of capacitors, whether they are the electrolytic variety or any other. Electrolytic capacitors are the only type that require polarity observation; all others may be connected without regard to polarity in the circuit.

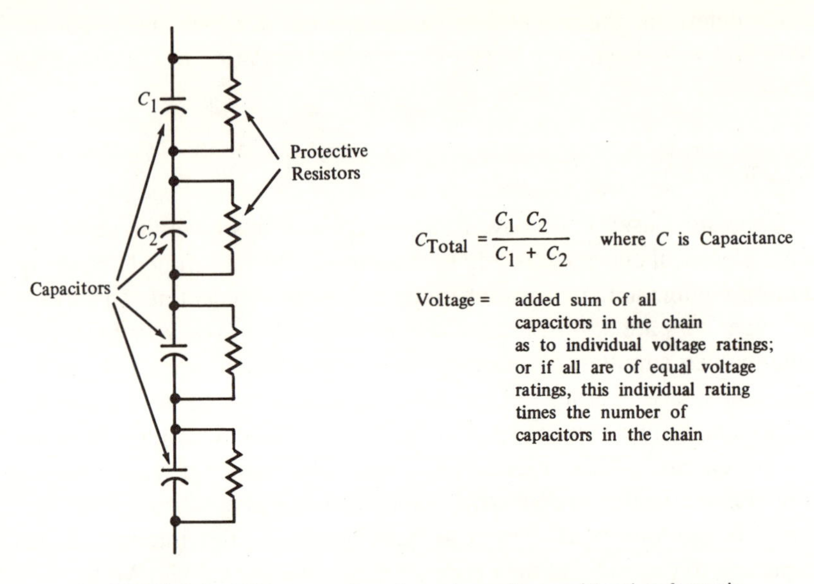

FIGURE 2-9. Filter capacitors in series with combination formula.

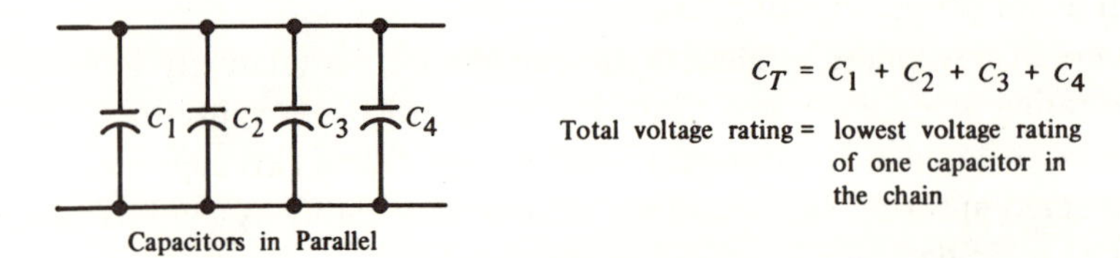

FIGURE 2-10. Capacitors in parallel with combination formula.

COMBINATIONS

Capacitors may be connected in series–parallel configurations in the same manner as solid-state rectifiers. Figure 2-11 gives an example of a possible circuit. Again, space limitations may make this type of circuit impractical for all but emergency use. It should also be remembered that only capacitors of the same manufacture and ratings should be used in any combination. Equalization problems result from mismatched capacitors in a series or parallel string, with one or more of the components possibly placed in conditions that will far exceed their ratings.

Series-connected capacitors should be provided with equalization resistors across each separate component to assure that each capacitor is matched and does not have to withstand the majority of the voltage from the rectifiers. Parallel-connected strings do not normally require this type of equalization, as the voltage rating is the same as for an individual component, and in every case the capacitance of each unit is added to the overall effective capacitance of the string as a whole.

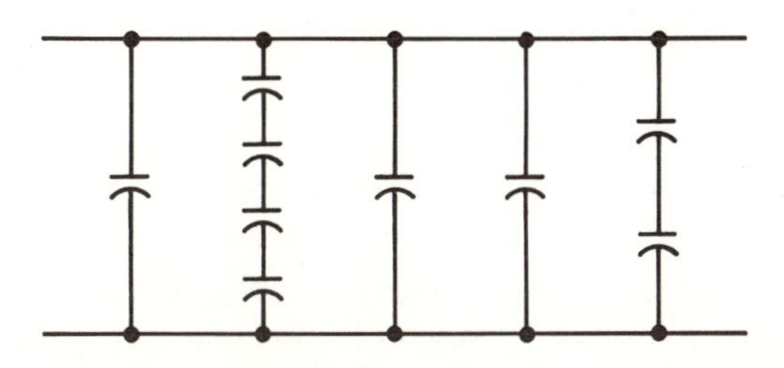

FIGURE 2-11. Series-parallel combination of capacitors.

Insulation

Capacitors connected in series may have to be insulated from the power-supply ground, because each capacitor may carry internal insulation that is inadequate to withstand the voltage difference between the top or start of the string and circuit ground. Figure 2-8 shows a string of eight capacitors rated at 400 V each with a total combined rating of 3,200 V (8 × 400 V). The top capacitor of this string could measure a voltage of 3,200 between its contact and ground if used to filter a voltage of this maximum value. The insulation between the internal voltage-handling portions and the case of the capacitor is designed to insulate voltages in the 400-V range. If the case of the capacitor is connected to the chassis of the power supply, which usually serves as the circuit ground, the insulation must then be effective for at least 3,200 V. A 400-V rating is inadequate and high-voltage arcing may occur, along with component damage and potentially lethal danger to any person close to the arcing. The capacitor is only withstanding or dropping one eighth of the total output voltage, but the potential voltage difference between it and ground is 3,200.

For operational and safety reasons, all series-connected capacitors should be insulated from circuit ground, and the metal cases of these components should be considered as potential high-voltage points in relationship to circuit ground.

Capacitors can be a safety hazard even after the alternating current has been switched off because of their ability to store current. Bleeder resistors are used in power-supply circuits for regulation and to bleed off the stored voltage. If the bleeder should fail, a potential high voltage could remain at the capacitor contacts. Shorting these contacts with an insulated screwdriver is the best procedure for safety when capacitors must be handled.

Chokes

Iron-core chokes are still used in some power-supply circuits. Ratings for these devices are expressed in henrys, millihenrys, or even microhenrys. As was learned earlier, a choke-input filter will tend to act as a capacitive-input filter unless the critical value of inductance is maintained. This value may be determined by the formula $L = E/I$, where L is the critical value desired, E is the output voltage of the power supply, and I is the current being drawn through the filter. Using this formula, the critical value of inductance will be expressed in henrys when current is expressed in milliamperes.

When the power supply is under no load, no current will be drawn through the filter. A critical value of inductance can only be maintained under loading conditions, as is stated by the formula. Thus, when using a choke filter, some current must be drawn at all times. The formula for minimum value of current is $I = E/L$, where L is the critical value of inductance in henrys, E is the output voltage from the power supply, and I is the unknown minimum current expressed in milliamperes.

This minimum current value is obtained when no load is being drawn from the power supply by the use of a bleeder resistor, which places a light load on the filter at all times. Adjustable resistors are often used to set this minimum load value.

BLEEDER RESISTORS

Bleeder resistors are placed across the output terminals of a dc power supply to discharge the power in the filter capacitors and to improve voltage regulation by providing a minimum load on the output. The value of resistance for the bleeder is determined by the output voltage. Correct values for voltage-regulating purposes will be discussed later, but for general purposes the value is noncritical and should range between 50 and 100 Ω of resistance per volt.

A much more important rating for the bleeder resistor is its power-handling capability expressed in watts. Using Ohm's law and a power supply with an output of 1,000 V, the power that must be dissipated or thrown off by the resistor may be calculated. Assuming a 100-Ω per volt resistance value for the bleeder, or 100,000 Ω, the Ohm's law derivative for determining current drawn through the resistor would be $I = E/R$, where I is the current drawn through the resistor in amperes, E is the output voltage (1,000 V) and R is the resistance of the bleeder (100,000 Ω). The correct answer is 0.01 A. Using the Ohm's law power formula of $P = IE$ and substituting 0.01 for I and 1,000 for E, the power in watts (P) that must be dissipated by the bleeder resistor is 10 W.

In practical use, this rating would be at least doubled for safety reasons, and a value of 25 W would most likely be chosen. A burned-out bleeder resistor is more dangerous than none at all, because the operator believes that he is protected from capacitor discharge. Inspection of bleeder resistors is done on a regular basis in most operations that use medium- and high-voltage power supplies. They tend to operate at high temperatures and have been known to fail completely.

Several resistors may be combined in series or parallel for a higher or lower ohmic value or for an increase in power-handling capability. Resistors in series add resistance, and resistors in parallel divide resistance, which is just the opposite of the capacitance formula for combining capacitors. Figure 2-12 shows the formulas for combining resistors. Power ratings always add, whether resistors are wired in series or parallel.

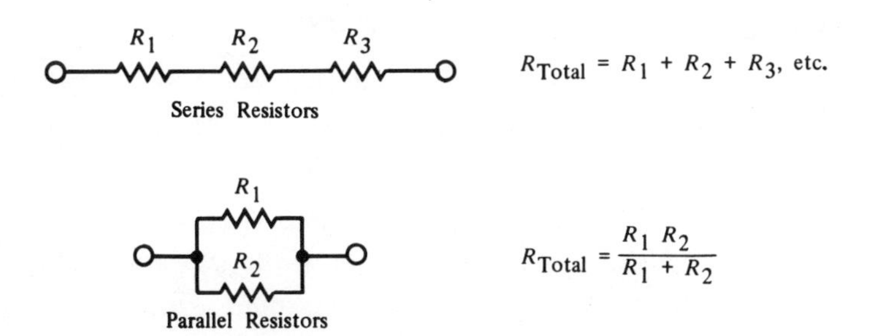

FIGURE 2-12. Resistors in series and in parallel with combination formulas.

After the correct value and power rating is found for the bleeder resistor, very high value, low power resistors are often placed in parallel with the main units. Should the main bleeder fail, these smaller resistors will slowly but surely drain the stored current from the filters. Values in the neighborhood of 1 megohm (MΩ) are satisfactory for these backup resistors, which have little effect on circuit operation while the main bleeder is functioning properly and need not be taken into consideration when choosing the values for the main bleeder unit.

QUESTIONS

1. Describe the construction of a typical power transformer.
2. With what kind of current does a power transformer operate properly?
3. What will result within the transformer windings if direct current is applied to the primary input?
4. What is the main function of a transformer used in dc power-supply construction?
5. What is the volt–ampere rating of a power transformer?
6. What factors determine the volt–ampere rating?
7. What is a CCS rating? What is an ICAS rating?
8. Define a rectifier.
9. Name the two types of rectifier construction materials.
10. What is a PRV or PIV rating?
11. Define I_{rep} ratings and compare them to I_s ratings.
12. List the advantages of solid-state rectifiers over vacuum-tube rectifiers.
13. What is a rectifier stack?
14. Explain rectifier thermal effects and the types of external protection.
15. Define transient voltage spikes and ways of preventing diode damage when they occur.
16. What methods are used to prevent surge current damage to solid-state rectifiers?
17. What should be done to equalize rectifiers wired in a series circuit? In a parallel circuit?

18. What function do capacitors serve in the filter circuit of a dc power supply?
19. What are the characteristics of electrolytic capacitors?
20. What is the total capacitance and voltage rating of two capacitors rated at 200 μF each and 300 V when wired in series? In parallel?
21. What safety procedures should be followed when wiring capacitors in series to form a high-voltage capacity bank?
22. What device is used in a dc power-supply circuit to maintain a minimum current drain on the filter choke?
23. Describe the two main functions of a bleeder resistor.
24. Should a bleeder resistor fail in the power supply circuit, what safety hazard could exist?
25. What safety precaution can be utilized in the circuit wiring of the bleeder resistor to assure proper bleed off of current?

chapter three

CONSTRUCTION TOOLS AND MEASURING INSTRUMENTS

The construction of dc power supplies generally requires a minimum of tools and accessories to arrive at a professionally finished product. By the same token, only a few electronic measuring instruments are required to check for proper operating parameters once construction is completed. Generally, power-supply circuits are far less complex in nature than are other types of electronic devices. Power-supply construction usually requires only good, basic building techniques throughout the construction procedure.

CHASSIS TYPES

Most modern power-supply circuits are constructed on an aluminum chassis that may be of any dimension, depending on the size of the power-supply components. Aluminum chassis material is normally used for ease of construction and good electrical properties, although steel is still used for very heavy power-supply components. A steel chassis is much more difficult to work with than one manufactured from aluminum and should be restricted only to applications where aluminum is not sufficiently sturdy to withstand the weight of the components.

Aluminum may be used in sheet form to build a chassis in the shop to conform to the desired dimensions for a specific project. Many electrical and electronic outlets stock chassis in various sizes and in many different thicknesses of aluminum for various construction purposes.

A chassis should be chosen to provide adequate mounting space for all power-supply components, while allowing for enough cooling space between the heat-dissipating parts of the supply. A power transformer, for example, should be mounted an inch or more away from other components to allow for some natural circulation of air for cooling purposes. Mounting configurations that are too cramped may cause component failure owing to an inability of the parts that heat up to dissipate their manufactured heat and also to components that become excessively hot from being mounted too close to other components that generate heat. A chassis that is larger than necessary may require a more expensive cover or cabinet, but it does leave room for additional modifications to the power supply at a future date or may allow for more than one power supply circuit to be mounted on the same chassis.

The best method to determine correct chassis dimensions is to set each of the power-supply components on a flat surface, observing correct spacing throughout. Once a comfortable arrangement has been arrived at, measurements of the space required to house these components can be taken. A commercially manufactured chassis that comes closest to these dimensions may be purchased, or a chassis conforming to the dimensions desired may be fabricated from aluminum sheeting material in the shop. The height of the chassis may be as little as 1 inch to as much as 4 or more inches and will be determined by the anticipated amount of components to be mounted below the surface of the chassis in the enclosed area. Large components are normally mounted on top of the chassis, whereas smaller components such as capacitors, some resistors, and meters are mounted below. For shielding purposes, most chassis are available with a metal plate that mounts to the bottom, completely enclosing the components within.

Once a chassis size has been chosen, the components are again arranged, this time on the chassis itself. This process assures that the chassis will be suitable for the components being used before any drilling is started. Many technicians will draw the outline of the component on the chassis, marking the areas where mounting holes must be drilled and other areas where circular holes must be cut. Once the desired mounting pattern and hole placement points are arrived at, all components are removed, and the drilling and cutting process can be completed.

To drill holes in aluminum, an electric drill motor is normally used with special metal-cutting drill bits. The majority of the drill bits that are used around the home and office are of the variety used to drill holes in wood. Metal-cutting bits should be used with aluminum because they cut a much neater hole without the problem of the bit traveling over the surface of the aluminum when the drill motor is first started. Most hardware and electrical supply stores carry metal-cutting bits in the various sizes needed. In an emergency, standard wood-cutting bits can be used, but the task will be much more difficult, and the finished work will probably not look as professional. After the drilling has been accomplished, the chassis should be turned on its back and any burrs of metal that remain around the holes should be removed with a rat-tail file and the sharp edges filed smooth.

If a construction project requires that large circular holes be punched in the aluminum chassis to accommodate specific components, this is best accomplished with a metal-cutting punch or die. These are often used to cut circular, square, and patterned holes in lightweight metals, and the finished result looks very professional. The chassis punch is used by drilling a small hole in the chassis at the absolute center of the intended hole. The bolt that draws the cutting portion of the punch into the mold is then slipped through this hole and the cutting portion screwed to the bottom. As this bolt is tightened, the cutting edge bores through the aluminum into the form or mold, and a perfect hole is cut every time. The main drawback to using cutting punches of this variety is the cost. Punches to cut holes 2 in. or more in diameter can be very expensive, although those that produce hole diameters of ½ and ¾ in. are usually reasonably priced.

Other types of metal-working tools may also be needed when building power supplies. Cutting dies that produce threaded holes in aluminum can create an even more professional project. Nibbling tools are also available that eat away at the aluminum when squeezed and, although tedious, unusually shaped holes may be formed in aluminum chassis in very short periods of time.

The more accurately the chassis is laid out as to component placement before the actual mounting is attempted, the easier the entire project will be to complete; a more professional appearance will also result.

After all holes have been drilled or punched in the chassis, rubber grommets should be mounted to prevent any cutting of wiring that may pass through these holes. Although the holes should have had all sharp edges rounded smooth, very small, sharp areas can still be present, which can eventually fray the insulation on conductors and cause short circuits during operation. Rubber grommets come in many standard sizes and resemble small rubber donuts. These grommets are fitted in each of the holes and protect the insulation on conductors with a protective rubber layer around the edge of the hole. Where very large holes are present, ones that are too large for any grommet size, the filed edges may be guarded by taping all conductors that pass through these openings. The conductors are carefully fitted through these holes and then wound with heavy electrical tape and secured so that very little movement is possible. Lack of movement will help to prevent the tape from sliding away from the portion of the wiring being protected as operation continues over extended periods of time.

Meters are mounted in the aluminum chassis by using metal punches similar to those described earlier. Punches are available in standard meter-mounting sizes of 2½ and 3 in. The hole is punched through the chassis, mounting holes are drilled in the four corners of the mounting area, and the meter is slipped through the hole and secured with metal nuts. Shielded meters require no special mounting arrangements; the metal shield simply slips over the back of the meter after it has been installed in the conventional manner.

Other holes may be required for specialized components and for mounting strips or terminal strips that are used to connect wiring at various points throughout the chassis. Small bolts are normally used to secure these strips, and they should be chosen in a size that will not allow movement after the nuts have been secured.

For specialized chassis cutting, heavy-duty tin snips may be successfully used after a primary slot or hole has been made. Snips are especially useful for making large square or rectangular cuts in aluminum, and specially shaped and formed holes are easily cut without excessive bending or distortion of the aluminum surface. Heavier-gauge aluminum chassis material will be best suited to this type of shaping.

Other specialized requirements may dictate the use of other types of cutting tools, but the ones discussed should be adequate for most needs and applications. Generally, whenever any cut is made in the aluminum chassis, regardless of the type of implement used, the edges should be filed smooth to prevent any sharp edges from interfering with installation of various hardware or cutting through conductor insulating materials. If this one rule is followed, most aluminum customizing work will go smoothly.

When warping or bending has accidentally occurred during a cutting or punching process, the material may sometimes be returned to a near-normal state by compressing the affected area in a bench vise. Pliers may also be used to reshape the aluminum. Incorrect drill bits, cutting punches, and other tools not intended for work on aluminum material are often the main cause of chassis damage. Choose the correct tools or make the proper modifications to existing tools before attempting any aluminum cutting work.

The last area regarding chassis and chassis materials to be discussed is that of bracing. Sometimes, when very heavy electrical components are used, the chassis will buckle or sag owing to the extreme weight. Heavy components should be mounted near the corners of an aluminum chassis wherever possible to take advantage of the added support that the chassis side panels offer. Where this type of mounting procedure is not applicable, some means of supporting the section of the chassis that experiences the most strain must be found. Lightweight aluminum rods are available that may be bolted beneath the chassis and secured to a bottom plate or the chassis sides, or they may be fabricated at a local metal shop. These shafts or rods should be sturdy enough to support the component that is to be mounted on the chassis, and they come in various gauges, weights, and dimensions. Those pieces intended to provide support between the chassis surface and the bottom plate will usually have flat ends; the type used for angle mounting between the chassis surface and the sides will have tapered or mitered edges. Depending on the component weight involved, several of these support rods, installed at strategic points throughout the chassis, may be necessary to give the required supporting strength.

Often supporting rods are considered to be a luxury item and are not used in situations where they are required. Slight sagging of the aluminum surface upon completion of construction may lead to severe buckling of the chassis as time passes. Metal fatigue may start, and the chassis may eventually have to be completely restored or replaced. Good construction techniques require the proper stressing of the chassis to assure continued, reliable service of the equipment involved. To do less than is required too often results in failure.

SOLDERING TECHNIQUES

Most contact wiring in modern power-supply design is secured by soldering. Printed circuit board installation and wiring are also accomplished through proper soldering technique. Although some high current primary wiring may be secured to the primary leads of the power transformer by special pressure contacts, almost 100 percent of the wiring done in power-supply circuits depends on solder joints for good electrical connections.

An explanation of proper soldering technique could fill a small text, but if certain basic concepts are followed, along with the proper soldering tools and accessories, very satisfactory results will be obtained in every case. Generally, there are two types of solder available on the modern market, resin core and acid core. Resin core is the *only* type of solder that can be used to form proper electrical connections. The acid-core solder should never be used in electrical or electronic soldering applications. Acid-core solder does not make a satisfactory electrical connection and usually presents a high-resistance solder joint when completed. The acid material in the solder will actually corrode circuit boards and components, completely ruining any type of electrical circuit.

Resin-core solder, the only type satisfactory for electronic work, forms an excellent electrical connection when properly applied, but should never be depended on to form a joint of any mechanical strength. In other words, if a contact must depend on the physical strength of the solder to maintain a connection, this joint will eventually weaken and fail. Solder should never be depended on to secure two conductors. A strong mechanical contact must first be formed by properly wrapping the conductors to each other or to an established contact. The solder is then applied to the mechanical joint to provide a lasting electrical connection. Should the mechanical connection originally formed be weak or inadequate, the conductors will gradually pull away, breaking the electrical connection. Improper soldering is responsible for a majority of problems associated with home-constructed electronic equipment.

Proper soldering technique involves, first, the establishment of a good mechanical joint. No movement of the elements to be soldered should occur even when the conductor or conductors are slightly tugged by hand. All contacts should be free of dirt, oil, or other foreign materials that will hinder good electrical contact. Once the mechanical rigidity of the joint has been established, the soldering iron or gun may be applied to the work surface. Apply this heating tool to the joint only, *not* to the solder. When the elements that form the joint become hot enough, the solder may be applied to the joint. The solder should melt and flow into the joint and remain in a liquid state for a second or so after the heating element is removed. Again, make certain that the joint reaches sufficient temperature to cause the solder to *flow* over the elements to be soldered. Do not drop random bits of solder onto the joint. This will result in an electrical malady known as a *cold-solder joint*. This type of defective connection will operate properly for a short period of time and then, unexpectedly, fail to provide proper electrical contact. This type of defect is very difficult to trace and correct once wiring has been completed and the equipment assembled.

When a correct solder joint has been completed, the solder around the joint will have a smooth and shiny surface and will completely cover the mechanical joint and flow into every part of it. Rough, dull solder surfaces indicate cold-solder joints that will soon fail, causing improper operation of the power supply.

Soldering iron or gun sizes vary and are chosen to accommodate the size of the job required. Small soldering irons in the 25- to 30-W range, often called pencil soldering irons, may be used for small, delicate work that may be found on printed circuit boards or in the rectifier section of a power supply where relatively small joints with small surface areas are found. Large irons or soldering guns with ratings of from 75 to 325 W are used for larger solder joints, which may be present at primary power leads, capacitor terminals, resistor contacts, and others. Solder is also available in several sizes and gauges that are designed for convenience on various-sized jobs.

Heat can be fatal to many types of electronic components. Rectifiers, transistors, integrated circuits, and other solid-state components may have a very low resistance to heat of the intensity required to form solder contacts. Using a soldering iron of too high a wattage when connecting these delicate components can cause permanent damage to their crystalline interiors. Low-wattage irons are most preferred for these types of soldering applications. As an added safety feature, it is desirable to provide some temporary means of preventing a lot of heat from traveling up the lead that is being soldered and into the solid-state material of the component. These heat sinks can be a pair of long-nosed pliers, which are used to squeeze the conductor being soldered at a point above the soldering area and near the point where this lead enters the component case. This method will allow proper soldering of the lead, and the pliers will form a protective heat sink, dissipating much of the heat before it can enter the component case. Special clips are available from electronic supply warehouses; they conveniently clip on various lead sizes and can be removed when all the soldering work has been completed. When soldering delicate solid-state components, many workers tend to hurry in order to keep the component heated for as short a time as possible. This is a good practice, but don't hurry to the point that an improperly soldered joint is the finished result. Take the time to complete the joint according to good soldering practices, but take no more time than is necessary.

Again, all the proper soldering techniques are completely useless if a good mechanical joint is not established before the soldering iron ever touches the work surface. Never suspend conductors or components for any great length between mechanical supports. Do not depend on solder to support even its own weight.

All solder joints should be inspected before final assembly of a power-supply circuit is attempted. Some joints that looked shiny and solid originally may look different after a few minutes. Any questionable joints should be reheated. Only a small amount of solder is necessary to form a good joint. Adding too much solder may decrease the effectiveness of a connection. Periodic inspection of solder joints should be done with electronic equipment. A regular maintenance plan assures continued operation.

BUILDING TOOLS

The assortment of tools required to build power-supply circuits is very fundamental. Most power supplies will require one large and one small flat-head screwdriver, a pair of long-nosed pliers, a medium-sized diagonal cutter, and, possibly, an adjustable wrench for bolting a cover onto the chassis. Large power supplies may require heavier tools than those specified, and some specialized circuits may require a few additional tools not mentioned here. By and large, the building of power-supply circuits can be accomplished with a normal complement of tools found in almost every shop or tool box. Other tools that may be helpful but are not absolutely necessary to complete construction would be a small- to medium-sized hammer, a set of nut drivers in various sizes, and a drill press.

A drill is the only electrically operated tool required for power-supply construction (other than the soldering gun); it is used to provide mounting holes in the chassis. Although a hand-operated drill will be adequate, an electric drill motor is much faster and, generally, more practical. A variable-speed model is desirable for working on aluminum chassis because the speed can be slowed until the hole is started and then speeded up once the bit is set.

Building accessories include a medium- to large-sized bench vise, assorted grommets, nuts, and bolts, and, possibly, some form of lettering material for the marking of switches and controls.

CHASSIS COVERS

It is essential that most power-supply circuits be completely covered with a metal case to prevent accidental contact with live circuits. These covers may be fabricated from aluminum in the shop or may be purchased already assembled or in kit form. Aluminum sheeting is available in do-it-yourself forms, and the perforated sheets are often used for covers on power supplies and other types of electronic equipment.

Constructing these covers is usually very simple. Measurements are taken of the length, width, and depth of the power-supply chassis. Another measurement is taken of the height that the components extend above the chassis top, allowing an extra inch or so for insulation and air-circulation purposes. The dimensions of the cover are easily obtained from these measurements, and the aluminum sheeting is bent to conform to these measurements. Perforated aluminum is often preferred for equipment covers because it offers the advantage of completely covering the electrical components while still allowing for a great deal of natural circulation through the perforated holes.

There are many types of covers and cabinets available on the modern market that are ideal for power-supply construction purposes. Some covers are very small and are made of plastic with a metal paneled front. These are ideal for small, low-voltage power supplies. Other models are ponderous and utilize heavy-duty construction; they are best suited to the heavy-current or high-voltage supplies, which generally weigh a considerable amount and place added stress on their housing compartments.

Many combinations are also available that offer a custom cabinet and a chassis of specified dimensions that conform to standard chassis sizes available as individual units. These combinations are ideal, because there is no problem in fitting the chassis to the cover or cabinet when construction is completed. For the neophyte builder, this offers an added psychological advantage for completion of the project. Many projects are started and nearly completed, but stay in an unfinished state because certain parts or components are not available or are too expensive. When a cabinet is all ready to house the finished product, an added impetus is provided to finish the project and to see it mounted in its shiny new cabinet, which has been staring the builder in the face since the start of construction.

Whatever type of cover or cabinet is chosen to house a power-supply circuit, it should conform to many standards, the most important of which is safety. Make certain that the circuit is completely covered and, generally, inaccessible to unknowledgeable persons. Compartment access doors and panels will be dictated by the placement of components on the chassis and by other factors, as will the mounting holes for meters and other controls that must pass through the front or side of the circuit cover. All controls should be insulated or grounded to prevent accidental shock due to a short circuit within the cabinet.

After the cabinet has been placed over the circuit, many builders will paint the finished product or finish it in a way to provide a professional appearance or to make it conform to other types of equipment already in use. Many products are available through electronic supply outlets for these purposes; they can be chosen for each specialized piece of equipment.

MEASURING INSTRUMENTS

Once power-supply construction is completed, some means of measuring performance is necessary. Even power supplies that provide metered outputs of voltage and current may need other measurements taken at the secondary of the power transformer, at the output of the rectifiers, or even at the primary of the power transformer.

A good vacuum-tube voltmeter or volt–ohmmeter of good quality is a necessary piece of equipment to any construction project involving electronic components. The output voltage can be checked with a voltmeter and the readings compared to those of the meter already provided within the power-supply circuitry. In this manner, the power-supply meter is calibrated against the external meter, which has a rated or known accuracy. Any variation in the two readings, if severe, can be corrected by changing the value of the meter multiplier resistor or any other voltage-determining part of the metering circuit.

Most volt–ohmmeters also contain a current scale that will read the load current drawn from the power supply when the test leads are connected in series with the output. Again, if a metered circuit is already provided, a reference is also provided for calibration purposes. If voltage output is high, the ac output from the transformer secondary can be checked by placing the meter function switch in the ac measurement position. If this value is also high, the primary voltage may be measured to check for a high input value. Without an external voltmeter of this type, these checks would be impossible and a cure for an apparent problem would have to be done by guesswork.

The values of bleeder and series resistors used in power-supply circuits may be checked with a volt–ohmmeter. Some resistors may not be close enough in value to their stated or printed resistance or ohmic figures to operate properly in a specific circuit. An accurate ohmmeter will allow the builder to choose resistors that have values within the proper operating tolerances required by a specific circuit. In many instances, the load resistance may also be measured, which will determine the amount of current demand.

The uses for a good volt–ohmmeter are practically endless for power-supply design, construction, and servicing. Several model types are available in the needle indicator and the digital read-out varieties. Good quality is always a prerequisite for these instruments. If reference resistors of a poor tolerance percentage are used in these devices, readings will not be as accurate as is required for many purposes. Generally, the higher the ohms-to-volt ratio, the more accurate the readings will be. An inexpensive, poor-quality meter can be used to give only a very rough indication of operation and, owing to inaccurate or incorrect readings, may be more of a hindrance than a help to power-supply check-out and operation.

Other electronic measuring devices of limited use in power-supply construction include ac ammeters, which are inductively coupled to the primary ac line to indicate alternating current drain. No physical connection is actually made, and most of these devices simply clip around the ac line cord. Sometimes a capacitance meter is also used to measure the capacitance value of filter components. Specialized power-supply designs may require other types of measuring devices, but for most purposes a volt–ohmmeter will be adequate.

Electronically regulated power supplies using transistors can usually be checked using the basic volt–ohmmeter circuit, but commercially manufactured transistor and diode checkers are available that will test these solid-state devices in seconds. These instruments can be very expensive if the automatic types are purchased. These pieces of equipment require no special test lead connections. The three leads are connected at random to the transistor leads and an immediate indication is given as to the quality of the transistor.

Special test equipment such as frequency counters may be required in dc-to-dc power supplies. These devices measure the frequency of the ac portions of the supplies. This is often a necessary measurement for dc-to-dc supplies and for dc-to-ac inverters that must power frequency-sensitive electronic equipment. The volt–ohmmeter is also necessary for the checks that must be made on dc-to-dc power supplies in order to determine the value of output and input voltage and current.

Safety is a prime consideration when checking out circuits in potentially lethal power supplies. Equipment that may not be operating properly is especially dangerous, because circuit inconsistencies can cause normally safe circuit points to carry a dangerous potential force. Some of the inexpensive volt–ohmmeters may be capable of measuring voltages with values as high as 5,000 V dc, but their test leads may not be insulated to this high value. With inadequate insulation, the danger of a severe electrical shock is always a possibility. This is another excellent reason for using only quality test instruments when attempting to determine the operating characteristics of electronic equipment such as dc power supplies.

QUESTIONS

1. Name the two types of metal that most power-supply chassis are constructed from. List the advantages and disadvantages of each.
2. What factors determine the dimensions of a chassis to be used for a power-supply circuit?
3. What steps should be taken to determine the correct size of a power-supply chassis?
4. What special requirements are placed on drill bits used for chassis work?
5. Describe the proper procedure for drilling holes in a chassis being prepared for power-supply construction.

6. What devices are often used to produce large circular or rectangular holes in an aluminum chassis?
7. What may be done to correct a warping or bending condition in a chassis caused by incorrect tools or improper cutting techniques?
8. Components that are very heavy should be mounted in what way on an aluminum chassis?
9. What may be done to a chassis to enable it to handle more weight when the use of very heavy components is contemplated?
10. What type of solder should always be used in electronic wiring?
11. What is a cold-solder joint? What are its causes?
12. Briefly describe proper soldering technique.
13. Before the soldering iron ever touches the joint, what requirement should be met?
14. What should be done to protect delicate, solid-state components from heat damage while soldering their connections to the rest of the circuit?
15. What is the prime consideration when choosing or building a chassis cover?
16. List some of the measuring instruments that are used to check the operation of power supplies and their components.
17. What is the prime consideration in choosing measuring devices?
18. What is an inductively coupled ac ammeter? How is it used?

chapter four

ELECTRONIC REGULATION

Voltage stability or regulation may be accomplished through the use of filter capacitors and chokes in several ways, but in some cases the amount of regulation is insufficient for many applications. Critical, solid-state circuitry often requires very efficient or *stiff* regulating circuits in their dc power supplies.

When high regulation is a necessity, electronic control of voltage stability is accomplished through the use of zener diodes, transistors, integrated circuits, or a combination of all three. A good electronic regulating circuit can maintain dc voltage output to within a few percent over large ranges of current demand.

ZENER DIODES

A zener diode is a special type of solid-state device that is used in the regulation circuitry of dc power supplies. This silicon junction diode presents an almost constant voltage drop at applied voltages that are greater than its breakdown point, which is the point between conductance and nonconductance. This point is sometimes referred to as the *zener knee*. All voltages past this point are effectively shorted to ground. This causes an increase in current drain, which is thrown off as heat from a series resistor within the voltage line at a point preceding the zener diode. Figure 4-1 shows a schematic of a low-voltage zener diode regulator circuit. Voltage ratings of zener diodes range from a fraction of a volt to several hundred volts, with power ratings, typically, to about 50 W. Physical mounting configurations may be in the form of a small, cylindrical case with wire leads, much like a resistor in appearance, or the case-mounted or *stud* variety, with provisions for transfer of heat to a metal mounting base for the high-wattage units.

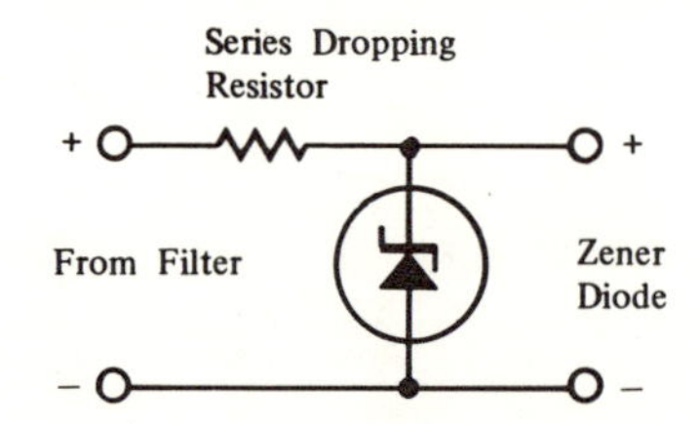

FIGURE 4-1. Simple zener diode regulator.

Almost all forms of electronic voltage regulation use a zener diode or the zener diode principle to perform their jobs of stabilizing dc voltage output under varying load conditions. Like the rectifier diodes discussed in earlier chapters, zener diodes may be placed in the regulator circuit in series and parallel configurations for increases in power-handling ability; however, the breakdown point will be altered and the components will exhibit different values when combined and operated as one unit. Two zener diodes in a series configuration will act as a unit with a 6-V breakdown point if each has a separate breakdown point of 3 V. A parallel configuration of the same two zener diodes would result in a breakdown point of approximately 1.5 V.

Zener diodes may be used in conjunction with resistors of low ohmic values to perform their regulating functions, or as a reference source for more highly sophisticated regulation circuits that incorporate transistors, thryistors, and other solid-state devices.

SOLID-STATE VOLTAGE REGULATORS

A solid-state voltage regulation circuit operates by comparing the difference between the proper supply voltage, which is determined usually by a zener diode, and the output voltage or a portion of it. The difference between the proper supply voltage and the actual dc output is instantly conducted in the form of an actuating or control signal to the other solid-state components in the regulator circuit. This control circuit can be thought of as a voltage control that will seek to maintain a zero difference in the two voltages (reference and actual). If a difference of only a small amount exists, a small amount of control or feedback signal will be conducted to the base of the control transistor(s), and the actual output voltage will be raised or lowered as is required to maintain a match with the desired voltage at the reference source.

Solid-state regulators, if properly designed, will maintain the actual output voltage of a dc supply to a value that is almost identical to the desired or reference voltage, even during periods of greatly varying current demands from the power supply. The term *constant voltage* is often used when referring to a source of regulated dc voltage. Circuits of this type are often used in critical electronic devices where an increase or decrease of only a fraction of a volt could result in improper overall operation. Electronic devices that generate radio frequencies are especially susceptible to voltage fluctuations, which can cause a change or drift in the correct frequency(s) and unsatisfactory operating traits.

Figure 4-2 is a simple schematic of a zener diode regulator circuit used with a conventional power-supply circuit. This is electronic regulation in its simplest form; it utilizes a series resistor in the dc positive line at a point in the circuit just prior to the zener diode, which is mounted across or shunted across the dc positive line to the negative line or ground. When the output voltage of the power supply exceeds the breakdown point of the zener diode, the diode conducts, grounding

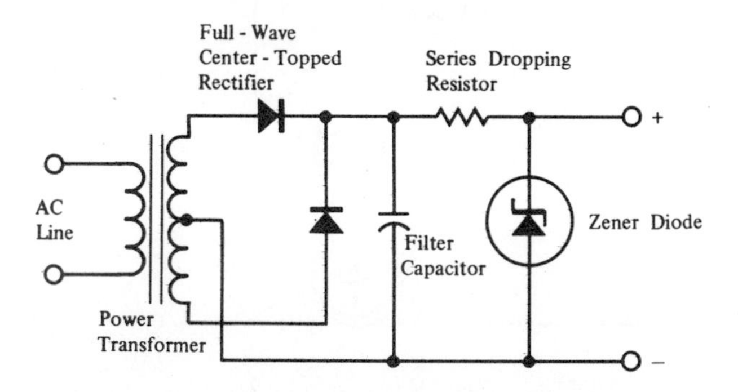

FIGURE 4-2. Zener-diode-regulated power supply.

out a portion of the dc voltage through the series resistor. This drain in direct current causes a voltage drop in the output of the power supply and lowers the voltage to a point that is equal to the breakdown point of the zener diode. Again, this is voltage regulation in its simplest form and is usually limited to devices that do not require too great an extent of regulation. Large variations in loading and current drain from this type of regulated supply will still cause fluctuations in the dc voltage output. These variations are greatly reduced when compared to those that would occur in a power supply that is unregulated electronically; however, the overall regulation of the zener diode circuit depicted in Figure 4-2 does not compare favorably with the more complex circuits that will be dealt with in this chapter, and which are much more comprehensive in design and operation.

This simple zener-diode-regulated supply finds major use in powering small electronic devices such as A.M. pocket receivers and the like, without the necessity of using small batteries.

SERIES REGULATORS

Figure 4-3 depicts a typical series regulator circuit. This type of regulator uses direct-coupled transistor amplifiers to amplify the error or difference signal that is derived from the zener diode reference source and a small portion of the output dc voltage.

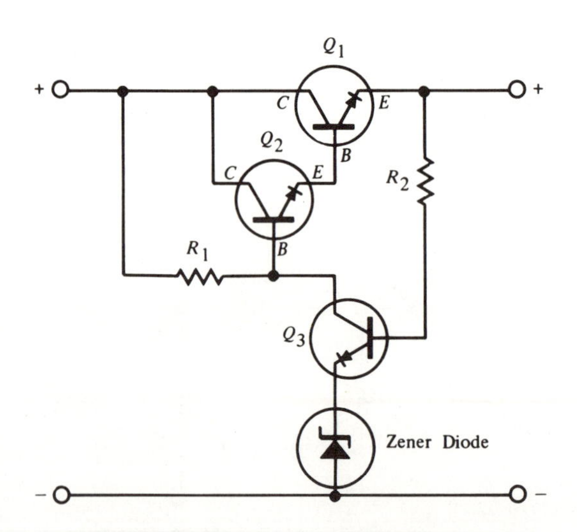

FIGURE 4-3. Series-pass regulator circuit.

By referring to Figure 4-3, it can be seen that the zener diode is placed in series with the emitter lead of transistor Q_3, which is the amplifier transistor. The difference between this diode breakdown point (the desired operating voltage) and the actual voltage output of the supply is amplified and is then fed to the regulating controls formed by Q_1 and Q_2. These two transistors act as a constantly varying resistor that maintains the amount of voltage appearing at the power-supply output at a constant level.

One problem often encountered in series regulator circuits like the one shown in Figure 4-3 is the current-handling abilities of the transistor or transistors that fall in series with the dc line. Since all current drawn from the power supply must pass through the transistor(s) that make up the control circuit, devices with the voltage and current ability to withstand the values placed upon them by the power supply must be used. Often the amount of current a solid-state device will safely handle is more of a limitation than is the voltage rating of the same device.

A different type of series regulator circuit is shown in Figure 4-4; it may solve the current-handling problem in many applications by reducing the dissipation factor or power-handling capabilities in the series transistor(s). In the device depicted in Figure 4-4, the maximum power that must be dissipated as heat in transistor Q_1 or Q_2 is greatly reduced when compared to the dissipation requirements placed upon the transis-

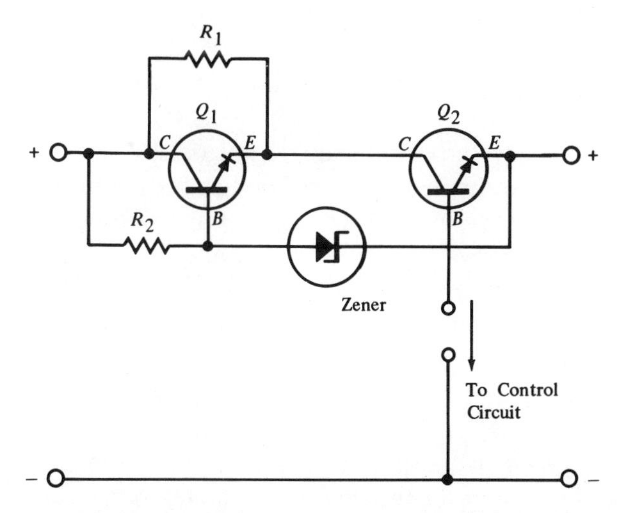

FIGURE 4-4. High-current series regulator.

tor in Figure 4-3. The dissipation requirements in the latter design may be reduced on the order of three fourths over the former regulator circuit. The remaining dissipation will occur in the resistor, which is shunted between the collector and the emitter leads of the control transistor.

Heat is a problem that is encountered in almost all electronic circuits. Transistors and other solid-state devices will fail during periods of excessive heating or temporary current overloads. The control of heating effects and dissipation is especially important when dealing with regulator circuitry of dc power supplies. In all but the very low current supplies using electronic regulation, the control transistors are mounted on heat sinks. In some applications, the heat sink will be the largest part of the supply, towering over the other components, often by a factor of three to one.

The series type of electronic regulator circuit is the most efficient regulator used in conventional supplies. It is also the most complex and expensive. For extremely critical applications, series regulation is almost always the most carefully designed section of the power supply.

SHUNT REGULATORS

When fairly good electronic regulation coupled with simplicity is a requirement in power-supply design, the shunt regulator circuit is often used. Shunt regulators are not as efficient regarding voltage regulation as are their series regulator counterparts, but they perform an adequate job while offering the advantages of simple design and relatively inexpensive components.

Figure 4-5 shows a typical shunt regulator; it can be seen that the two transistors are shunted across the dc line between the positive and negative leads, as opposed to the series connection of the series regulator circuit mentioned earlier. The control transistors do not have to withstand the full flow of current being drawn from the power supply and can be of a smaller, less expensive variety. Current that passes through the shunt elements, transistors Q_1 and Q_2, varies with changes in the current drawn from the supply. This variation is reflected across the series resistor R_1, which is connected in series with the positive lead and the load of the power supply so that the output voltage is maintained at an almost constant level. Again, a zener diode is used in this circuit as a reference voltage for the transistors to act upon. The

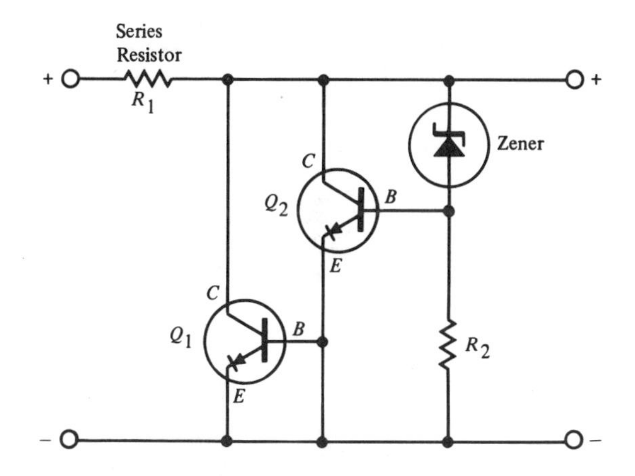

FIGURE 4-5. Shunt regulator circuit.

remaining resistor R_2 merely sets up proper operating conditions for transistor Q_2 and will usually be of the low-wattage, high-resistance variety. R_1 must be rated to handle the current and dissipation that will be required of it in each particular circuit as it passes the full value of power-supply current demanded by the load.

VARIABLE-VOLTAGE REGULATORS

The voltage regulator circuits previously discussed have the advantage of maintaining the dc output voltage from the power supply to a value of within a fraction of 1 percent of optimum. One disadvantage of this type of regulated supply is the inability to alter the voltage output to power other devices that may require slightly different voltage values for proper operation.

A variable-voltage regulated power supply is easily built for many applications with only a few minor changes to circuitry and the addition of a few electronic components. Figure 4-6 shows a very simple series regulator circuit that uses one control transistor and a zener diode. The resistor sets the proper operating bias on the base of the control transistor.

Figure 4-7 shows the same voltage regulator circuit with variable-control circuitry added. As can be seen, an additional transistor and three more resistors are all that is required to convert the original circuit to allow for control of output voltage. In the circuit of Figure 4-7,

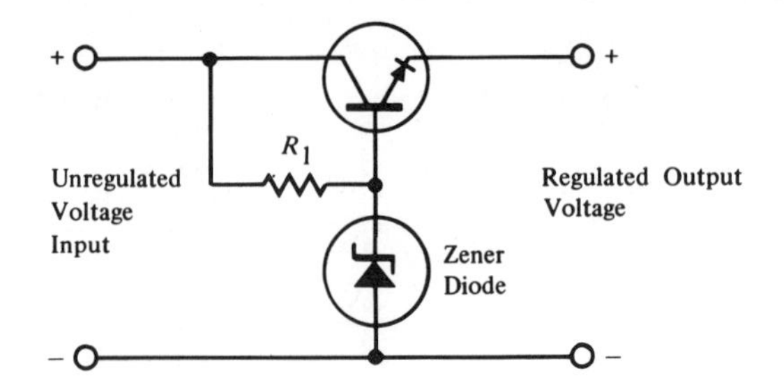

FIGURE 4-6. Simple one-transistor series regulator.

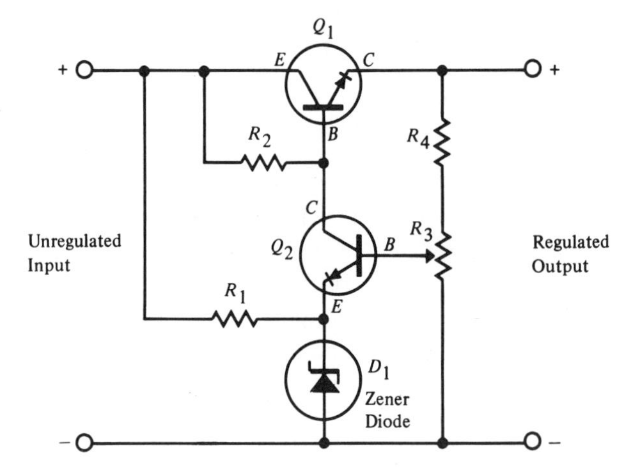

FIGURE 4-7. Series-pass regulator with output voltage control.

the lowest output voltage that may be obtained from this supply will be the breakdown voltage of the zener diode D_1, plus the small amount of voltage that is present between the base and emitter of transistor Q_2. Resistor R_1 provides bias voltage to the zener diode, which provides the reference voltage to be compared to that of the variable resistor or potentiometer R_3. Highest output voltage is dependent upon the type of rectifier circuit used and the lowest unregulated voltage input to the regulator circuit.

CURRENT REGULATORS

Although it performs a different function than that of a voltage regulator, a current regulator or current limiter is often incorporated into

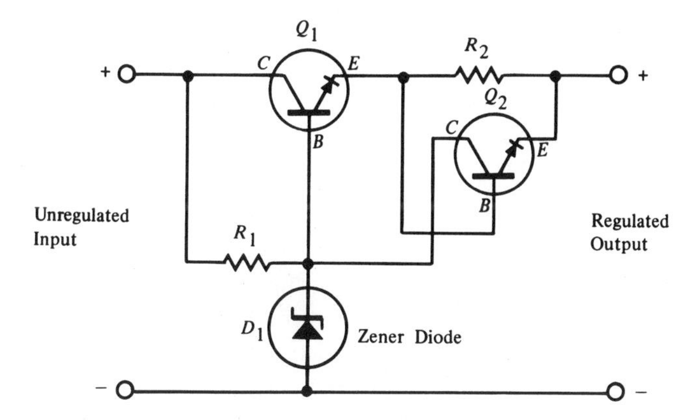

FIGURE 4-8. Current-limited voltage regulator circuit.

voltage regulator circuits and deserves discussion in this part of the text. Current regulators are used mainly to limit the amount of current that is allowed to pass on to the voltage regulator circuit and the load. This protects the components in the latter stages of the power supply and all circuits in the load. Current is not regulated in a direct fashion as is voltage in the voltage regulator circuits, but rather in an indirect manner by regulating voltage that is developed by the load current across a resistor that lies in series with the power-supply output and the load.

Figure 4-8 depicts schematically a current-limited voltage regulator. The circuit is very similar to the two previous schematics discussed and includes the basic one-transistor series regulator circuit. Transistor Q_2 and resistor R_2 comprise the current regulator portion of the circuit, with R_2 placed in series with the supply output and the load. As the output current begins to increase, the current regulator transistor Q_2 conducts more, dropping the base current on the voltage control transistor Q_1 which conducts less and effectively limits the output current. The output current is limited to a point that is equal to the voltage between the base and emitter of Q_2 divided by the value of limiting resistor R_2.

To provide a constant current output, the voltage across the series resistor R_2 must also be held at a constant level. Proper resistor size is important in a circuit of this type, because when excessive heating occurs, resistors will change their values, which will also change the current. A resistor that is twice as large as is required by the Ohm's law power formula should be used to maintain as uniform a regulator as possible.

It can be seen from studying Figure 4-8 and reading the text that current limiters are directly dependent on a constant voltage, which must be maintained by the voltage regulator, and are in themselves a type of *voltage* regulator used to indirectly control the current output of a regulated power supply.

Current regulators have wide use in many electronic fields, including everything from hobby electronics to laboratory equipment. Fuses do not usually open the circuit in time to prevent damage to solid-state components during a current overload. A current regulator greatly reduces the chances of harming components in delicate electronic equipment, and many times current-limiting devices are designed to sense an overcurrent situation and automatically shut down the primary voltage to the supply until the overcurrent condition ceases. The supply can then be reactivated without the necessity of replacing fuses or other electronic parts.

SWITCHING REGULATORS

Voltage regulator circuits are usually grouped into either series (or series-pass) and shunt types, but another type, the switching regulator, is used in circuits that require an unusually high amount of regulation proficiency.

Switching regulator circuits have a distinct advantage over the other types, especially in applications that require a very high amount of current output. Series regulators subject their series control transistors to high amounts of power dissipation, which causes heating in the components. In a switching regulator, the control transistor is either fully conducting or completely off, as opposed to the series regulator in which the transistor is usually partially conducting at all times, a condition that creates the high dissipation requirements. Switching regulators usually exhibit a low dissipation requirement for their control elements, along with a very low forward voltage drop through these transistors.

Transistors or thryistors may be used to make up the control elements of switching regulators, with transistors usually preferred for most dc-to-dc regulating purposes. Thyristors are more suited as ac-to-dc devices in switching regulators and where unusually high values

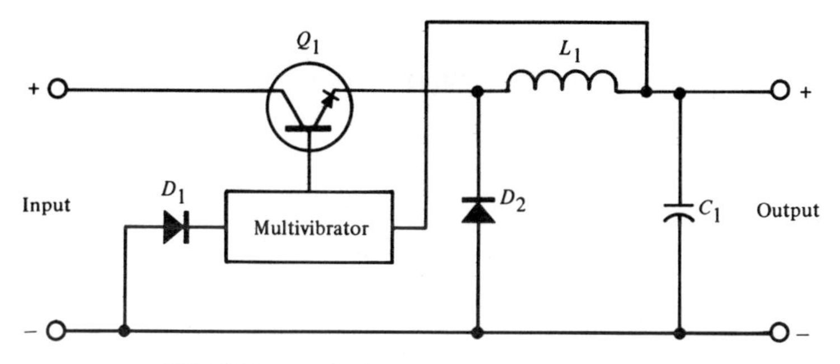

FIGURE 4-9. Switching regulator circuit.

of voltage and current are passed within the power supply. Switching regulators require a high amount of filtering of the output, and sometimes additional regulating circuits to keep the output ripple at a low value. These circuits often serve as the input to additional stages of standard series regulators for an extremely efficient or stiff source of dc voltage and current.

Figure 4-9 is a schematic of a switching regulator circuit using only a few components, as compared to the more elaborate series regulator circuits previously discussed. The reference diode D_1 determines the amount of on and off time of the multivibrator circuit, which is used to switch the control transistor full on or full off in short, fraction-of-a-second pulses. When the transistor is full on, the load derives its current directly from the voltage source. When the multivibrator circuit switches the control transistor full off, the current is supplied through the energy that has been stored in the storage circuit made up of L_1 and C_1. Diode D_2 conducts when the transistor has been turned off and activates the storage circuits. By disregarding all components lying to the left of and including D_2 on the schematic, a familiar circuit can be identified, a standard choke input filter, which was discussed in detail in Chapter 2.

AC-TO-DC REGULATORS USING SILICON-CONTROLLED RECTIFIERS

One shortcoming of transistors used for voltage regulation purposes is their cost for high-current, high-voltage power supplies. A more economical solid-state device for these types of applications is the silicon-controlled rectifier (SCR).

Silicon-controlled rectifiers may be used to regulate the output voltage at a point in the power-supply circuitry that follows transformation, rectification, and filtering, as is accomplished with the regulator circuits already discussed; but SCR devices may also be used in a manner that accomplishes the rectification, regulation, and filtering all in one combined circuit, since the SCR automatically rectifies any alternating current that passes through it. These devices are available with PRV ratings in excess of 1,000 V and with current ratings of 100 A and more, and at a fraction of the cost of high-voltage, high-current transistors.

Silicon-controlled rectifier voltage regulator circuits accomplish their task by controlling the phase of the ac waveform, adjusting the portion of the ac wave during which conduction occurs. Like the switching regulator control transistor, the SCR is either full on or full off, and efficiency is very high while heat dissipation is kept at a low value. The SCR switches off electronically each time the ac wave drops to zero and stays in this off state for a few millionths of a second. A separate circuit, usually consisting of a low-powered transistor, is used to switch the SCR to the on mode by sensing the polarity of the ac wave. In this manner, the SCR is turned on during each half-cycle (plus and minus). This sensing circuit must read the waveform properly so that the SCR is turned on in a way that will maintain the value of output voltage during large variations of the input voltage. With the sensing circuit acting during each half-cycle, the overall unit is able to make almost instant corrections during periods of sudden input voltage changes by altering the rate at which the SCR control device conducts.

The electronic on–off switching of the SCR will usually generate harmonics or noise of the type encountered in audio devices in the home when a light switch is turned on or off. A filtering circuit consisting of a choke and capacitor or capacitors is usually necessary to circumvent the bad side effects of using this type of regulator circuit.

SUMMARY

The basics of power-supply voltage and current regulation have been dealt with in this portion of the text. Regulator circuits tend to be the most complicated part of most power supplies, and as new devices become available on the commercial market, new circuits and circuit principles are developed overnight. The regulators discussed here are, for the most part, of a relatively simple nature, and, as with all types of electronic circuits, more complicated arrangements can always be broken down into many simple circuits just like the basic ones discussed here.

Voltage regulator design will be determined by efficiency required, expense, and the operating parameters. Devices used to construct these regulator circuits will be determined in the same manner. With the basic knowledge obtained from this chapter, the reader should be able to determine the type of regulator circuit required for a given set of parameters. For instance, if good regulation is required of a power supply that must deliver 40 V at a constant current drain of 1 A, a shunt type of regulator may be the best choice while still maintaining a low cost figure. If the output of the supply were to be 40 V at from 0 to 2 A, varying, a series regulator circuit would probably be called for. If the output were to be 400 V at 50 A, an SCR regulator would be the most practical. By knowing the basics of these regulators, a specific design can be arrived at and built with only a few minutes of planning and forethought in many instances.

QUESTIONS

1. Define the term *breakdown point* when applied to zener diodes.
2. In what voltage and power ratings are zener diodes usually found?
3. List the solid-state devices that are often used to form regulating circuits in power supplies.
4. What purpose does the zener diode serve in most voltage regulator circuits?
5. What does a voltage regulator do to the final output voltage of a power supply?
6. What problem(s) can a poorly regulated power supply cause in other electronic circuitry?
7. Describe the characteristics of a series regulator circuit.
8. Describe the characteristics of a shunt regulator circuit.
9. List any advantage (s) of a series regulator over a shunt regulator circuit.
10. List the advantage(s) of a shunt regulator over a series regulator circuit.

11. What considerations are required of solid-state devices, especially transistors, that are used as control elements in high-current power-supply circuits? What special mounting requirements are involved?
12. What advantages are to be found in a variable regulated voltage supply?
13. What is the main purpose of a current regulator?
14. What advantages are there in using power supplies with current regulator circuits?
15. What is the main advantage of a switching regulator over a series or shunt regulator?
16. What operating conditions decrease the dissipation of the control transistor in a switching regulator circuit?
17. What type of application would be most suited to a transistor-type switching regulator? To a thyristor-type circuit?
18. What is the main advantage of using thyristors in voltage regulator circuits?
19. What conditions dictate the use of SCR devices in regulator circuits?
20. Why is a filtering circuit usually necessary with SCR regulators?

chapter five

VOLTAGE MULTIPLICATION

When ac voltages are rectified and filtered using conventional methods, an increase in output when compared to the measured ac input voltage is accomplished under light loading conditions. The peak output from a half-wave rectifier with a capacitive-input filter is about 1.4 times the rms or measured ac average voltage across the secondary of the supply transformer winding. If the rms voltage is measured as 10 V across this secondary, the dc output, after half-wave rectification and filtering with a capacitive-input filter have taken place, will be measured at about 14 V under conditions of little or no current demand or loading. This peak voltage rating is the maximum voltage that may be obtained with the ac input stipulated. The same voltage peak of 1.4 times the rms voltage holds true for the full-wave center-tapped and the full-wave bridge rectifier circuits as well.

By using conventional methods of capacitive-input filtering and rectification, a voltage multiplication of sorts has taken place, with the output direct current at about 1.4 times the ac input.

It is also possible to multiply voltages to even greater values by combining rectifier stacks into arrangements with capacitor filters to form a series circuit that will produce two, three, four, five, and more times the dc output that is obtained from any of the conventional rectifier and filter circuits already discussed.

Voltage-multiplying circuits were first developed in the days when vacuum-tube rectifiers were the only means of changing alternating current into direct current. Vacuum tubes are not really practical for these circuits because many of the rectifiers and capacitors operate at different voltage potentials, and separate filament supplies would be required for the many tubes used. Even at the comparatively low cost of transformers during that period when compared to today's prices, voltage multipliers were expensive items and were not widely used.

Today, with silicon rectifiers requiring no external power supply, the voltage multiplier has come into heavy use in almost every field of electronics. A transformer with a secondary voltage of 10 V rms can now deliver the usual 14 V when rectified conventionally or 28 V in a voltage doubler. The term *doubler* is applied to the peak dc voltage obtainable from a conventional rectifier assembly and not to the rms ac value. A voltage-doubler circuit will deliver almost *three* times the rms transformer value in dc output voltage. By using the voltage-doubler circuit with the 10-V transformer, a 28-V dc output power supply may be constructed with little or no extra expense over that which would be incurred by building a standard 14-V dc power supply. In many circuits, the only added component would be another filter capacitor. By using voltage-multiplying circuits, less expensive and readily available power transformers may be used to obtain many different dc output voltages, depending on the circuit configurations used.

With voltage multiplication we do not get something for nothing. If a power transformer is designed to safely deliver 1 A of direct current using conventional rectification, when the voltage is doubled, half of this power value, or 500 mA, is all that may be safely drawn from the power supply. The transformer will still be delivering its rated 1 A of current to the circuit when only ½ A is being drawn from the supply as a whole. Ohm's law, $P = IE$, will bear this out; power is equal to the output voltage times the current drawn. For example, a transformer with a 10-V secondary rms reading will supply about 12 V under a moderate load from a conventional rectifier assembly drawing about 500 mA. The power drawn from this transformer will be equal to 12 times 0.5, or 6 W. If the voltage-doubler circuit is used, the supply will deliver about 24 V under a current drain of 250 mA (0.25 A). Using the power formula, 24 times 0.25 equals, again, 6 W. It can be seen that for both supplies the power drawn from the transformer is the same. The voltage can be doubled, tripled, or multiplied even more, but with each multiplication step in voltage, an equal division step in current must be provided for. Voltage can be multiplied, but maximum output power cannot be increased using any type of rectification and filtering circuit.

Voltage doublers are widely used in high-voltage power supplies. A transformer that delivers a secondary rms voltage of about 6,000 V requires an unusually large number of windings on the secondary core; and when it is used in a standard full-wave center-tapped rectifier configuration with an appropriate filter circuit, it will deliver about 3,000 V dc under load. The same dc output voltage may be obtained through voltage-doubling circuits by using a 1,100-V rms transformer, which is smaller, uses less turns in the secondary, and is less expensive. A voltage-doubling circuit will require only an additional capacitor or capacitor bank, as compared to the full-wave center-tapped circuit, and the overall size of the power supply will probably be more compact and lighter in weight. Remember, though, that if the safe current demand on the transformer is 1 A, the safe dc output current from the voltage-doubler circuit can be no more than one half of that amount, and in practical use it will be a little less than half owing to ohmic losses in the rectifier and filtering circuits.

Voltage-multiplying circuits are able to multiply the rms voltages at the input by rectifying these values and storing portions of them in capacitors that operate in a store–release manner during different portions of the ac cycle. The stored potential is added to that which is being derived directly from the voltage source one time for a doubler circuit, and several times for the more advanced multiplying chains of three times rms and more.

To maintain good regulation, high values of capacitance are required, because the capacitors are wired in series, and capacitors in series have decreased capacitance values while providing increased voltage-rating values. Thus, in a voltage-doubler circuit using two capacitors valued at 10 μF each, the total capacitance of the circuit is only 5 μF. This applies to voltage multpliers of the full-wave variety, which act during both halves of the ac cycle. Full-wave multipliers are the only practical type of voltage-multiplying circuits used in modern devices. Half-wave multpliers act on only one half of the ac cycle and present large regulation problems. It may seem unusual, but all multiplier circuits are made from single half-wave rectifiers, just as the full-wave center-tapped rectifier circuit is made up of two half-wave circuits in parallel acting on each half of the ac waveform.

VOLTAGE DOUBLERS

Figure 5-1 is the schematic of a full-wave voltage-doubler circuit. Diode D_1 charges capacitor C_1 during one half of the ac cycle. During the next half-cycle, D_2 charges the second capacitor, C_2. These capacitors are charged to 1.4 times the rms value, but the load represented by the resistor to the right of the circuit draws its power from the capacitors connected in series. Each capacitor is charged to 1.4 times the rms, and the load receives direct current at 2.8 times the rms voltage input. One capacitor receives current from the transformer during each half-cycle, so full-wave operation is obtained. The ripple frequency is 120 Hz, as opposed to the 60 Hz that would be obtained with half-wave operation.

Examining the circuit in Figure 5-1, we see that by removing D_2 and C_2 and connecting the load to the top and bottom of C_1, a standard half-wave rectifier is formed. By removing the top filter capacitor and diode, C_1 and D_1, another half-wave rectifier is formed, but one that acts on the other half of the ac cycle. Earlier in this text it was stated that all complex circuits were simply combinations of the most basic circuits; the full-wave voltage-doubler is a prime example of the truth of this statement.

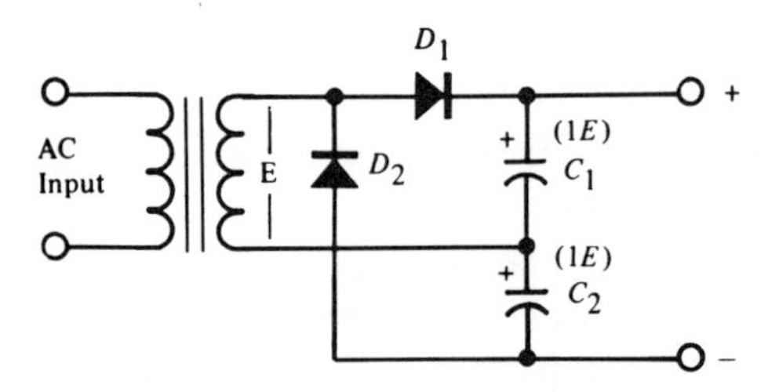

FIGURE 5-1. Full-wave voltage doubler.

Referring again to Figure 5-1, the series-connected capacitors carry the full voltage potential. C_1 carries half the value of the output voltage, or about 1.4 times the rms value, while C_2 carries an almost identical value. In practical application, resistors of a relatively high ohmic value are placed across each capacitor to assure that each carries the same amount of voltage. Small differences in capacitance and internal resistance will cause an unequal split if these resistors are not included in the circuitry. In cases where the two capacitors exhibit large differences in construction and capacitance value, one may receive a portion of the voltage that is large enough to exceed its safe operating parameters, and failure could occur. If the output of the power supply is 100 V dc, each capacitor should be rated for at least 50 V dc, with a 25 percent higher voltage rating being preferred to be certain of staying within the operating limits of the component.

Capacitors of the same capacitance value *and* made by the same manufacturer are usually used to assure as close a match in operating characteristics as is possible. The equalizing resistors can only do so much in matching capacitors, and can make up for only very minor differences between the two components. Equalizing resistors usually serve two purposes in a voltage-doubler supply, (1) to equalize the resistance of capacitors in series, and (2) to form the bleeder chain. It will be noted that, when equalizing resistors are used, the resistors themselves are actually wired in series, and the overall resistance value is multiplied by the number of resistors used. Two 20,000-Ω resistors in series, each rated at 10 W, form a larger resistor of 40,000 Ω with a total power rating of 20 W.

VOLTAGE TRIPLERS

Figure 5-2 shows a voltage-tripler circuit of the full-wave variety. It can be seen that this is nothing more than a voltage-doubler circuit with an additional half-wave circuit added. The voltage tripler delivers three times the peak dc voltage available from a standard rectifier assembly and 4.2 times the rms value under low loading conditions (3 × 1.4). Three capacitors are used to charge the direct current, which is

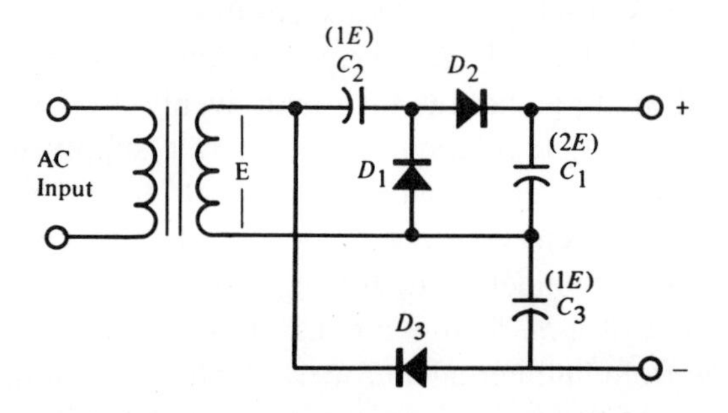

FIGURE 5-2. Full-wave voltage-tripler circuit.

supplied by the three rectifier diodes. Diodes D_1 and D_2 form a voltage-doubler circuit, but in this configuration it is a half-wave voltage doubler, acting on only one half of the ac wave. Diode D_3 charges C_3 during one portion of the ac cycle to a value of 1.4 times the rms voltage value. D_1 and D_2 plus C_2 charge capacitor C_1 during the other half-cycle to a value of 2.8 times the rms voltage.

This requires further explanation. D_1 and D_2 form a half-wave voltage doubler. This circuit has components that conduct during each half of the ac cycle, but it is still a half-wave doubler. During the first portion of the cycle, D_1 conducts and charges C_2 to 1.4 times the rms value. During the other half-cycle, D_2 conducts, placing C_2 and the transformer voltage in series. The output capacitor C_1 is now charged up to the combined voltage of capacitor C_2 and the transformer voltage, or 2.8 times rms. C_1 is now in series with C_3, which carries half the stored voltage of C_1, and the resulting output voltage is three times that obtained from a conventional rectifier circuit. (This explanation is most difficult to follow without referring to Figure 5-2.) Again, D_1, D_2, and C_2 charge the top output capacitor to 2.8 times the rms voltage. This is accomplished during both halves of the cycle. One half charges C_2; the other half charges C_1 with the stored current from C_2 and from the transformer's rectified current. C_3 has been charged during one half-cycle as well, and finally adds its value to the value stored in C_1 to obtain the desired output of the tripler circuit.

To make explanation even more difficult, a voltage tripler of the type just described is a full-wave device, but is unbalanced and will exhibit a 60-Hz ripple in the output. The imbalance occurs because the voltage pulse to the load is twice as large during one half-cycle as it is during the other. The difference in the voltage charge delivered to the two output capacitors creates this condition. This 60-Hz ripple will be present in all full-wave multipliers of the types discussed that use odd multiples of the input voltage. Higher values of capacitance should be used with these odd-multiple circuits than with evenly divisible circuits to provide good voltage regulation. Remember, a supply with 120-Hz ripple is more easily filtered than one with a 60-Hz ripple.

The voltage-tripler circuit is one of the most difficult concepts in voltage multplication to understand. It is a *full-wave* circuit that exhibits the ripple output of a *half-wave* circuit. It derives a great deal of its output voltage from an internal *half-wave* doubler circuit that acts during both halves of the ac cycle in the manner of a *full-wave* rectifier circuit. By understanding this circuit and the voltage doubler discussed earlier, the entire principle of voltage multipliers to much higher multiples of the transformer voltage is readily understood.

VOLTAGE QUADRUPLERS

Figure 5-3 shows a voltage-quadrupler circuit of the full-wave variety. Again, half-wave doublers are used, two of them, to produce a combined output of four times the peak dc voltage, or 5.6 times the rms value. D_1 and D_2 form the first half-wave doubler, charging C_1 to 2.8 times the rms value during one half of the ac cycle. D_3 and D_4 perform the same function during the other half-cycle, charging C_2 to the same potential. C_3 and C_4 come into play during the opposite portion of the cycle, which charges their respective output capacitors, as was explained during the discussion on voltage triplers. Unlike the tripler circuit, the voltage quadrupler receives charging pulses of equal value to each of its two output capacitors during the full cycle, and a balanced output is obtained, exhibiting a standard full-wave 120Hz ripple.

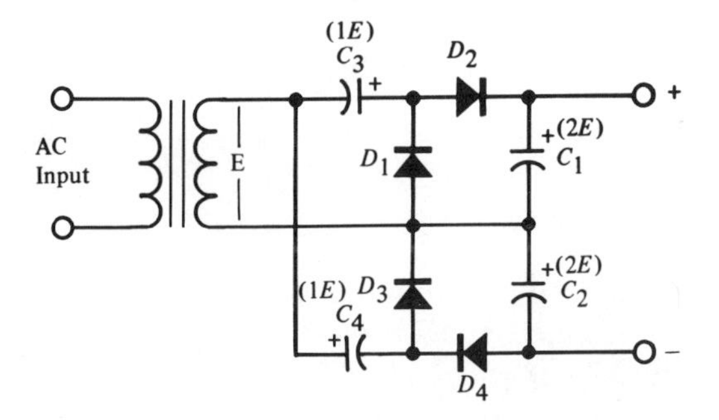

FIGURE 5-3. Full-wave voltage quadrupler.

The capacitor designations of 1E, 2E, and so on, indicate the amount of voltage each capacitor is charged to, as well as the voltage rating which that capacitor must be able to safely withstand. E is the peak dc value or 1.4 times the rms value.

By continuing the process of adding half-wave rectifiers and half-wave doubler circuits, the voltage-multiplication chain can be built to deliver almost any multiple of the input voltage that is desired. Remember that, when using a voltage doubler, tripler, quadrupler, or even quintupler, the available output current must be divided by two, three, four, five, or the equivalent of the figure by which the voltage is multiplied.

HALF-WAVE VOLTAGE MULTIPLIERS

Half-wave voltage multipliers have limited use in modern electronic design. Although half-wave doubler circuits are integrated into full-wave multipliers, discrete circuits are seen only in limited quantities, and then are usually relegated to powering small electronic hobby devices and transformerless radios. The latter use can be hazardous when medium to high output voltages are obtained from the multiplier, because one side of the half-wave multiplier circuit is at a full ground potential.

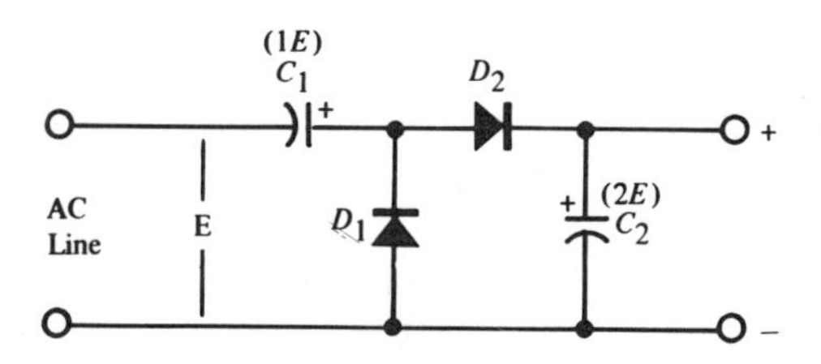

FIGURE 5-4. Half-wave transformerless voltage doubler.

Figure 5-4 shows a transformerless voltage-doubler circuit. This circuit uses no transformer, but derives its ac input directly from the 110 V ac line. Many ac appliances constructed of metal are grounded to the ac ground point. If the proper polarity is not observed and the ground side of the line is connected to the top lead or plus side of the doubler line, every grounded appliance becomes a potential safety hazard if the doubler circuit is contained in a metal box with one side of the doubler output grounded to this case. Anyone coming in contact with this box and a metal appliance at the same time could receive a severe electrical shock, not of the normal 110 V but of the full dc output of the doubler, which would be somewhere in the range of 280 V or more if even higher multiples were used in the multiplier circuit. A transformerless supply must include a 110-V male plug that is wired so that the hot side of the ac input is always connected to the top of the doubler, and the cold or ground wire is always at the bottom of the multiplier (as shown schematically).

Even with a polarized plug, accidents can occur, especially if the multiplier happens to be an octupler (eight times the peak voltage); here a potential of over 1,200 V is present at the output terminals of the supply or between the metal case of the supply and the ac ground if proper wiring precautions have not been taken.

Transformerless supplies are almost universally used in devices that are completely surrounded by nonconductive, plastic cases, such as inexpensive table radios. As discrete circuits, they have little use in modern electronics design.

Full-wave multipliers may also be operated directly from the ac line, but, owing to a vast difference in wiring configuration, no portion of the ac line is ever at ground potential. These circuits are much safer for transformerless operation and possess better regulation characteristics as well. Half-wave doublers are notorious for their poor regulation factors.

Referring back to Figure 5-4, capacitor C_1 is rated at E peak or at least 1.4 times the rms, which would be about 170 V when using standard house current. C_2 is the output capacitor and is rated at twice the voltage of C_1.

Figure 5-5 shows several schematics of a voltage tripler, quadrupler, and even sextupler. All the half-wave circuits shown in Figure 5-5 are used in conjunction with a standard power transformer, which provides isolation from the ac line and a much higher level of user safety.

COMPONENT RATINGS

Generally, the component values of devices used to form voltage-multiplier circuits will be the same as those required for other types of power-supply circuits. The voltage and current drain will dictate the PRV and current ratings of the diodes used, just as the voltage in each of the discrete circuits will dictate the voltage rating of the capacitors. Bleeder resistors will be dependent on the load resistance derived from the current and voltage at the output terminals. Capacitor value in microfarads should be a bit larger than would normally be used to maintain a good level of regulation when using standard rectifiers.

To find the correct minimum capacitance required, the load resistance is first determined by dividing the output voltage by the value of

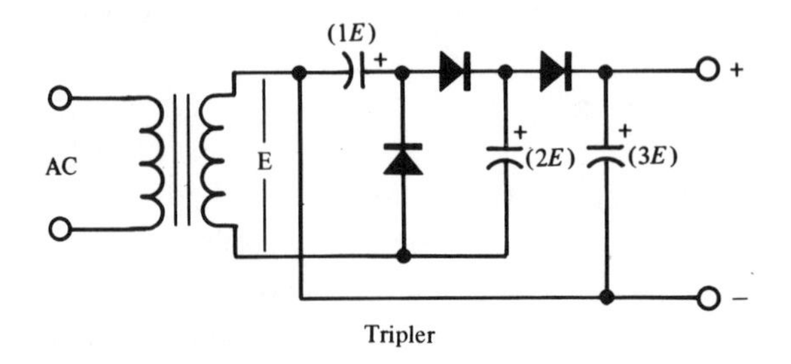

Tripler

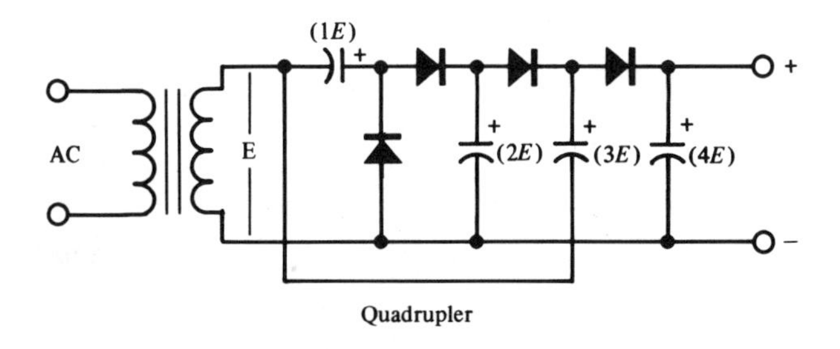

Quadrupler

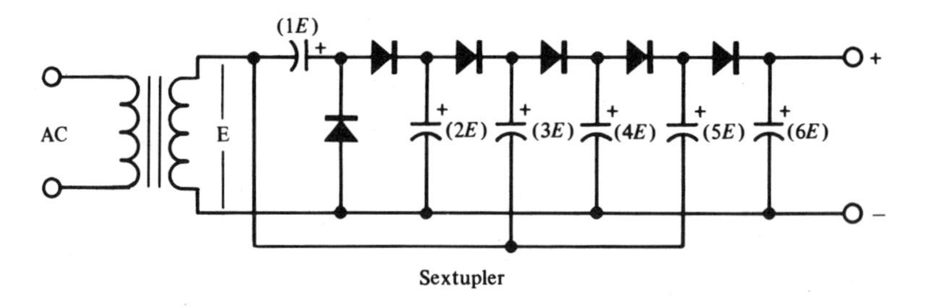

Sextupler

FIGURE 5-5. Other multiples of half-wave circuits.

current drain expressed in volts and amperes, respectively. Use the chart in Figure 5-6 to then determine the correct capacitance. This chart is read directly for a full-wave voltage-doubler circuit only. For a full-wave tripler circuit, multiply the capacitance value obtained from the chart by 1.5. For a quadrupler, multiply the capacitance value from the chart by 2, and so on. For half-wave multiplier circuits, the same chart will give the correct values of capacitance by figuring the load resistance, using the chart to derive a capacitance value, and multiplying that value by the multiple of the multiplier circuit. A doubler will take the chart capacitance and multiply that value by 2, a tripler by 3, a

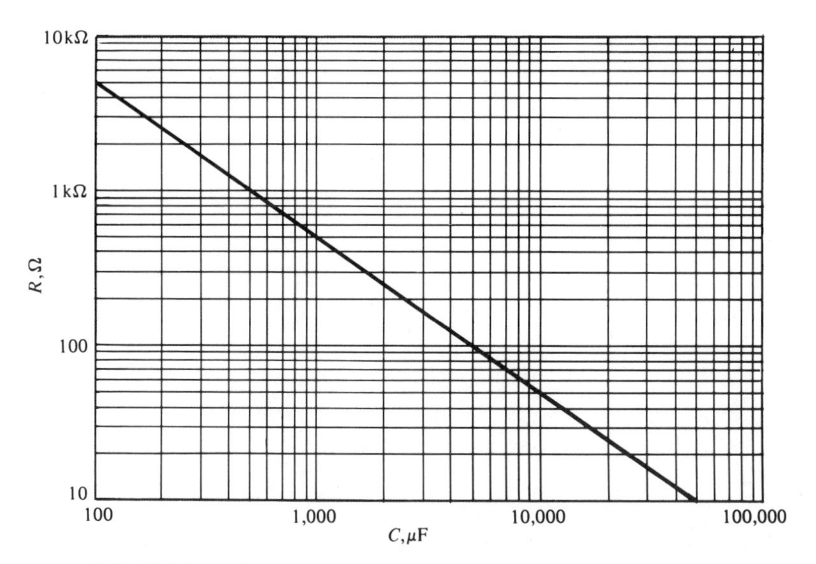

FIGURE 5-6. Capacitance chart for voltage-multiplier circuits.

quadrupler by 4. This method applies to determining the capacitance values for half-wave multipliers only.

Observe voltage ratings when choosing capacitors. Use a capacitor with at least the peak voltage rating indicated in each particular circuit. Since the voltage multipliers discussed all use capacitor-input filters, voltage and current surges will be present as with the standard capacitor-input circuitry studied earlier.

OUTPUT VOLTAGE

We have learned that under no load the output from voltage multipliers will be 1.4 times the rms voltage times the multiple factor of the circuit (doubler, tripler, quadrupler, etc.); but when these supplies are placed under loading conditions, a voltage drop will occur, owing to circuit resistance, to approximately 1.25 times the rms voltage. Another voltage drop that is not included in the under-load formula, but which does present itself in the circuit, is the voltage drop across each diode used in the rectifier assembly. Most silicon diodes will drop about 7/10 V. If the multiplier circuits are used to develop dc voltage of a low value, these voltage drops across the rectifiers may amount to a considerable percentage of the overall output, but most practical applications of voltage multipliers can allow for a variance of a few volts. All the formulas discussed in this section are based on proper sized capacitors being used, as determined by the formula and chart provided in the previous section on component sizes.

REGULATION

Using the capacitor chart for proper capacitance size, adequate voltage regulation will be provided using the multiplier circuits discussed. Full-wave multipliers maintain about the same 1 percent regulation as can be anticipated in other nonmultiplying circuits. Even half-wave circuits maintain their regulation to a level that is comparable to other half-wave supplies. Again, full-wave circuits are the only type of voltage multipliers that should be used to supply power to most modern devices. Half-wave multipliers should be used only where voltage and current demand are very noncritical, and then only with a transformer to isolate the circuitry from the ac ground.

As with other rectifier–filter circuits, voltage multipliers can be used with many of the electronic regulator circuits discussed in Chapter 4 for a higher percentage of voltage stabilization. Regulation may be improved slightly in marginal cases by increasing the capacitance of the output capacitors; but past a certain point, a law of diminishing returns is established and no noticeable improvement in regulation will be noted. Also, a simple electronic regulator will probably be less expensive to add to a multiplier circuit than the replacement of each capacitor with a larger unit, especially when medium to high voltages are produced by that supply.

VOLTAGE-MULTIPLIER USES

Other than the obvious use as a main power supply for assorted electronic equipment, the voltage multiplier may be called on for many different applications. Many power transformers are still made with 5-V windings at the secondary in addition to a medium- to high-voltage winding and a 6- or 12-V winding. The 5-V winding was used heavily in the days of the vacuum-tube rectifier. Many of these rectifiers required a 5-V filament supply at several amperes. In many applications today, this 5-V winding is simply not connected to the rest of the power-supply circuit and goes unnoticed and unused. But by using this winding in conjunction with a voltage-doubler circuit, a dc voltage source of approximately 12½ V, under load, is available for very little added expense. This extra voltage source can be used to control relays, to provide a voltage source for other solid-state equipment used with the devices getting their power from the main portion of the supply, or even as an alternative filament supply for equipment requiring such. These uses can be greatly expanded by adding a simple voltage and current regulator.

Voltage multipliers may also be used with nonmultiplying rectifier circuits to provide a means of decreasing and increasing the power output of transmitting devices. Where it is inconvenient or impossible to lower the ac primary input to a transformer to control transmitters, a voltage-doubler circuit may be switched in and out of the secondary winding leads of the supply transformer. Figure 5-7 shows a possible circuit. With a transformer secondary rms voltage of 500 V, the dc output voltage will be approximately 650–700 V using the full-wave bridge circuit; but if higher voltage is needed for a power increase, the voltage-doubler circuit may be switched in. The dc output voltage will

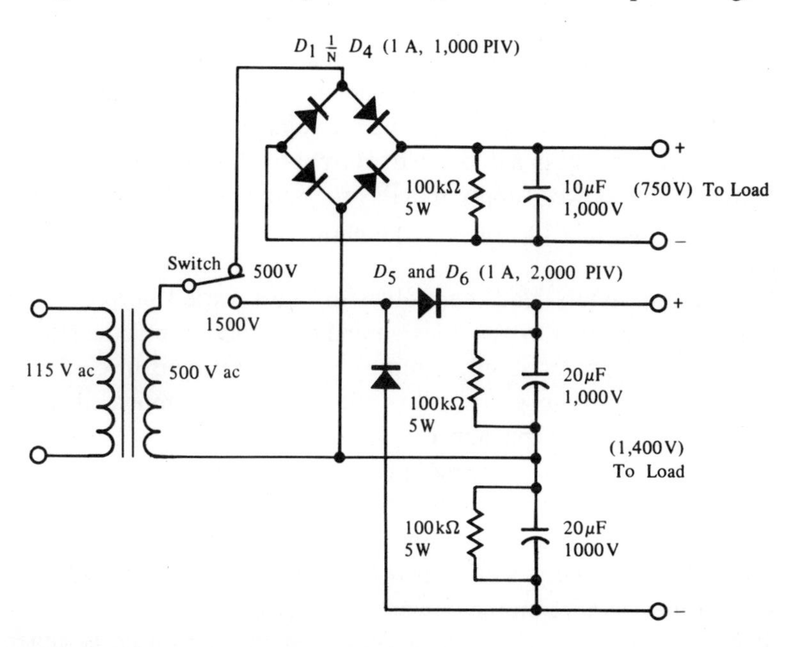

FIGURE 5-7. Bivoltage power supply using a standard as well as a voltage-doubler rectifier configuration.

now be in the neighborhood of 1,300–1,400 V. If this power supply provides voltage and current to the plates of a tube-type transmitter, the voltage-doubler circuit may be used as an effective means of doubling the power output of the transmitter by doubling the plate voltage.

Another method of controlling power output and class of operation in tube-type transmitting circuits is by changing the amount of voltage on the screen grid of the output tube(s). A voltage multiplier could be called into service to raise and lower this value. A 12-V winding on a power transformer can be pressed into service as a 30-V power supply for the control grids of many popular transmitting tubes.

Voltage-multiplier circuits add a great deal of versatility to already existing power supplies and allow a great deal of specialization when designing and building new power supplies. With enough planning, almost any voltage may be derived from the standard ac sources available at the secondary of many power-supply transformers.

SUMMARY

The advantages and disadvantages of the various voltage-multiplier circuits have been dealt with in this chapter. Voltage multipliers do not give the builder something for nothing. A voltage multiplication factor of 2 will always result in a division of current by 2, so the power output delivered by the power supply remains the same as with conventional rectifier circuits; only the value of potential force or voltage has been changed.

Voltage multipliers lend themselves to many varied purposes, but they are not a shortcut to good design and construction practices when dealing with dc power-supply circuitry. The same care in choice of components is mandatory, as well as slightly more complicated calculations when figuring capacitance.

Voltage multipliers require more components than most of the standard rectifier configurations, but the relative simplicity of design and monetary savings over a power transformer of increased size and voltage output are usually well worth the extra components and extra space that they require on the power-supply chassis.

Since voltage doublers, triplers, and so on are often used to generate high voltages from medium-voltage transformers, the safety factor should again be mentioned. When dealing with potentials of 2,000,

3,000 and even higher voltages, there is no such thing as a slight electrical shock. These are lethal voltages and should be treated as such.

QUESTIONS

1. In a voltage-doubler circuit operating from a power transformer that delivers 280 V ac rms, what will the dc output be under conditions of no load?
2. A voltage tripler that delivers 630 V dc at the output under no loading conditions derives its input to the rectifiers from an ac source of what rms voltage?
3. A full-wave bridge rectifier and capacitive-input filter deliver 100 V dc at the output under no load and with a current potential of 1 A. If the same ac source were used with a voltage doubler, what would the output voltage and safe current rating be?
4. List the main cost advantage of a voltage-multiplier circuit over a standard rectifier circuit when both are designed to deliver the same dc output.
5. What frequency is typical of a full-wave voltage-doubler circuit regarding the ac ripple component in the output?
6. What is the ac ripple component in a full-wave voltage tripler?
7. Why do voltage multipliers that multiply by even multiples have a higher ripple component in the output than those that have odd multiplying factors?
8. Describe the electrical processes that take place within a voltage doubler of the full-wave variety.
9. Describe the processes that accomplish an output of three times the peak ac input in a full-wave voltage tripler.
10. Voltage multipliers that produce four or more times the peak ac value at the dc output are made up of strings of what type of electronic circuits, connected in aiding configurations?
11. What danger is there in operating a half-wave multiplier circuit directly from the ac house current source?
12. Where are half-wave multipliers mostly used in modern design?
13. A multiplier circuit that delivers 150 V under no-load conditions will deliver how much voltage when placed under a medium load?

14. Output capacitors in voltage multipliers are wired in what configuration?
15. In a voltage-doubler circuit, the two output capacitors are individually rated at 40 μF. What is the total capacitance of the output section?

chapter six

METERING

Once a power-supply design is arrived at and construction is completed, some means of measuring the voltage and current output is often necessary. Meters are usually connected to the output of the power supply to constantly monitor the parameters of the circuitry. The readings supplied by these instruments serve several functions in addition to determining the output voltage and current of the supply. These readings may also indicate the performance, to a degree, of the electronic apparatus receiving power from the supply. The input power of transmitting devices is often indicated by observing the voltage and current measurements at the power supply. By using the Ohm's law formula, *power* equals *current* multiplied by *voltage,* the total consumption of the electronic device can be accurately derived. If a metering circuit is also used in the output network of the transmitting device, the comparison of the power-supply input reading and the power output reading can determine the overall efficiency of the transmitter. Most transmitters, especially the higher-powered varieties, require accurate metering of power input to comply with regulations governing their operation. A poorly constructed metering circuit could actually result in illegal operation.

Power-supply metering circuits are also used to indicate malfunctions of the equipment being powered. A sudden large current drain could indicate a short circuit at some point in the wiring. When the equipment being powered is removed from the supply, the readings can be rechecked. If the high-current indication is still present, the problem will probably lie within the power-supply circuitry; but if the indication shows little or no current drain, a normal condition for power supplies with no load across the output, the problem will lie within the device being powered, and steps can be taken to correct this condition.

In modern power-supply design, metering circuits deserve as much attention to detail as do the rest of the circuits that comprise the overall unit. Meters range in quality and price. The more critical power supplies will require more sensitive metering circuits. Power supplies used with home entertainment equipment and other noncritical electronic devices can be a little less accurate in many instances and, therefore, less expensive.

METER MOVEMENTS

When an electrical current flows through a conductor, a magnetic field is formed around the conductor. The strength of the magnetic field is in proportion to the amount of current flowing. When the conductor is looped or formed into a coil, the magnetic field is increased in strength. The amount of increase or the multiplication factor is determined by the number of turns and the diameter and turns spacing of the coil. The magnetic field is actually a miniature reproduction of the magnetic field that encompasses the earth; both have a negative and a positive force, a north and south pole.

Opposite forces attract each other; like forces repel. This is dramatically demonstrated by taking two magnets and holding them close together. One side of the magnets will attract; reversing one magnet results in a repelling force. It is upon these forces that modern meter movements are based; the needle or indicator that is usually attached to the coil is pulled or pushed, depending on the polarity of the magnetic fields present within the coil, which also moves with the indicator and reacts with a permanent magnet that is stationary within the meter case.

Figure 6-1 shows a basic d'Arsonval meter movement, named after Jacques d'Arsonval, who first used the magnetic field principle to construct a meter of this sort in the year 1882.

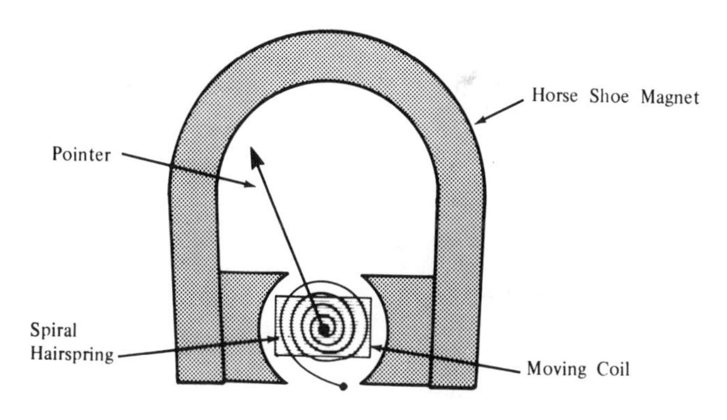

FIGURE 6-1. D'Arsonval meter movement.

The magnetic field produced by the permanent magnet reacts with the field produced within the moving coil, which is determined by the current flow within that coil. The attraction of the moving coil to the magnet is opposed by the spiral hairspring, which causes the coil and the attached indicator to stop its movement at a point where the attraction of the magnet is equal to the opposing force in the spring. The current flow is then indicated by the pointer's position against the backdrop of a calibrated scale.

Meter sensitivity is determined by the strength of the field produced by the magnet as well as by the number of turns in the moving coil and the opposing strength of the small spring. Theory of operation calls for little else in the meter movement, but actual construction is quite a different matter. The moving coil must be suspended in some way, and it is within this suspension system that a tiny amount of friction will exist to further impede the movement of the coil. Modern meters have very negligible friction factors, and in most applications the friction component can be disregarded. Only when very small amounts of current must be accurately indicated does this opposing force become a problem because of its higher ratio with the current induced within the meter movement.

Figure 6-2 shows another meter movement. This detailed drawing shows how the moving coil is normally suspended within the meter case for proper operation. Meter movements are very delicate devices and may malfunction owing to dust and other foreign particles entering the case. Meter repair is usually accomplished with jeweler's tools and by individuals with formal training in working on miniature devices. Although removal of dust can sometimes be done by removing the case

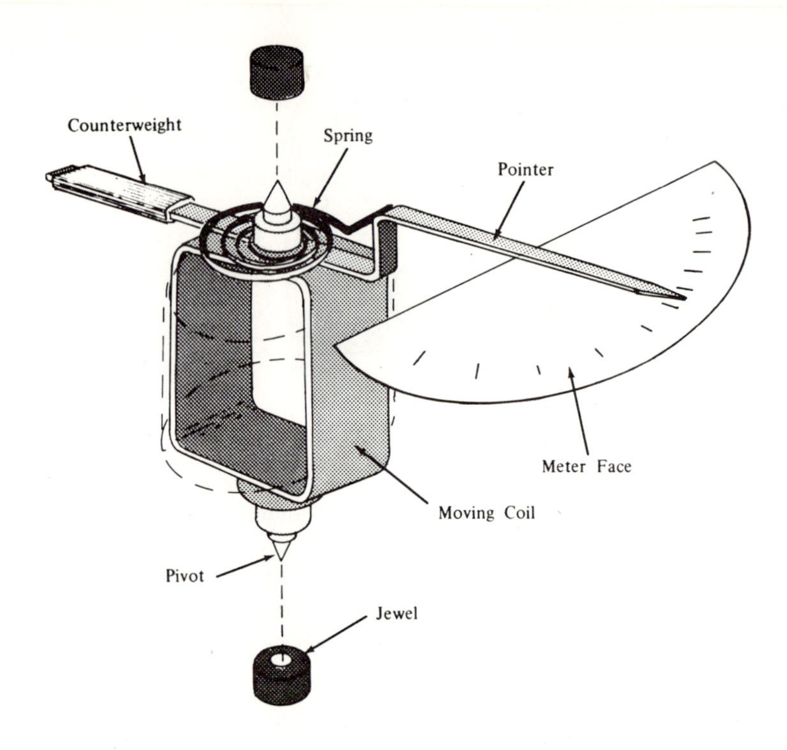

FIGURE 6-2. Coil suspension of d'Arsonval movement.

and gently blowing on the movement, this method is not recommended, because out of the case the movement is subject to even greater damage. The hairspring is designed to provide a specific tension, and even slight bending by an untrained finger can cause complete inoperation of the entire meter movement.

Meter casings vary greatly as to attractiveness and style, and utility and mode of operation dictate different configurations. Some meters are designated for mounting in a nonmetallic cabinet; others may be used with metal or nonconducting mounting surfaces. Mounting limitations, if any, are usually printed on the meter case or included with the operating instructions. Some meters are shielded to prevent stray magnetic waves from interfering with overall accuracy. This shield usually takes the form of an external metal case completely surrounding the movement, but others may be constructed with internal shielding that is not immediately obvious upon initial inspection.

Most meters achieve maximum accuracy when operated with the

scale in a flat or horizontal position. Although in most applications, the scale is placed in a vertical position, horizontal mounting usually places the movement pivots in a vertical position that assures rotation around a positive point on their bearings. If this ideal mounting condition is not practical, vertical-scale mounting is usually acceptable for almost every application except where extreme accuracy is required.

Meters used to monitor operating conditions and values of medium-to high-voltage power supplies should be considered as voltage points and potentially dangerous areas of construction. Proper spacing of cables and general insulation practices should be adhered to, as when attempting any construction of power supplies or other devices bordering on the high-voltage scale of voltage values. A broken scale plate could bring the pointer or indicator needle within possible human contact, causing a severe electrical shock.

VOLTAGE MEASUREMENT

Modern meters are current-sensing devices that act upon the current flowing within the meter and its associated circuitry. Voltage readings are possible with these current-sensitive devices by utilizing the Ohm's law formula, which states that the circuit voltage may be determined when the values of circuit current and resistance are known. The formula $E = IR$ states that voltage (E) is equal to the current (I) times resistance (R). This formula may be transposed in several different ways to find unknown values of voltage, current, or resistance when two of these values are already known. For instance, if the current and circuit voltage are already known factors, resistance may be arrived at by using the variation of the Ohm's law formula for resistance determination, $R = E/I$. This means that circuit resistance R is equal to the voltage divided by the current.

To properly measure voltage with a d'Arsonval meter, a current-sensitive measuring device, a resistor is placed in series with the meter coil so that a current that is proportional to the voltage is allowed to pass through the movement. For example, a meter movement will read full scale when 1mA of current is passing through its coil. If a voltage-measuring capability of 100 V is desired from this meter, a proper series-resistor value must be arrived at to pass 1 mA at 100 V for a full-scale reading. Using the Ohm's law formula for resistance determination, $R = E/I$, and substituting the known values of voltage and current, we obtain $R = 100/0.001$ (0.001 A is equivalent to 1 mA). The final answer to this equation is $R = 100{,}000\ \Omega$. A total resistance in the metering circuit of 100,000 Ω or 100 kΩ is required. This does not mean that an additional 100,000-Ω resistor is added to the metering circuit, however, because there is already a quantity of resistance within the meter coil that must be taken into consideration when accurate measurements are a must. The resistance of the moving coil within the meter is subtracted from the resistance required to properly meter this circuit, and the remaining figure will dictate the correct value of the resistor. If the resistance of the moving coil within the meter is 100 Ω, a value of 99,900 Ω is required for the extra resistance.

In practical application, the resistance of the meter coil is often ignored, because the slight difference in required resistance within the circuit does not alter the reading obtained on the meter scale by a great degree. Factory-built meters that are designed to read circuit voltage usually contain specially made resistors of high accuracy, depending on the quality of workmanship. The more accurate meters have resistors with a very low error tolerance; cheaper models contain less expensive resistors.

With the voltage-measuring meter circuit complete as described, when a voltage flows within the metering circuit, a certain amount of current is passed through the resistor and the moving coil, causing a corresponding reading of current to occur on the meter scale. When the circuit voltage is equal to a value of 100 V, the corresponding current through the meter coil will be 1 mA or full scale on the meter face. A 100 V scale voltmeter has been constructed from a 1-mA milliameter.

Manufactured meters, of course, have had all the computations and constructions already completed to arrive at a properly functioning meter; but it is good to know how to construct a voltmeter from a meter designed only for current monitoring, because in many applications construction of specialized meters is required. The most practical means of accomplishing this is to start with a high-quality microammeter or milliammeter and design the movement to pass the required current to match the voltage ranges within the circuit. Many commercial broadcast metering requirements dictate that all applicable readings must be taken from a certain portion of the scale on the meter face where readings tend to be more accurate. Most meters read more accurately when the indicator is in a mid- to full-range position rather than in the low-range part of the scale. A specialized meter may be designed to have a scale that requires the indicator to advance to at least midscale for the lowest reading that is likely to be encountered. Readings taken from the last half of the meter scale are generally more accurate than those taken from the first half of the same scale.

This type of design flexibility can often only be obtained through the modification of manufactured meters and meter accessories. It is even possible to alter an ac voltmeter to indicate current, although this necessitates the removal of the internal resistor. The altering of milliammeters for voltage-measuring purposes is, however, much more prevalent and practical. In many instances, it is not even necessary to alter the current scale already on the meter face. For example, a microammeter that was designed at the factory to read from 0 to 50 μA could be set up to read from 0 to 5, 0 to 50, 0 to 500, and so on, volts, and the present scale could be used directly or by mentally adding a zero or two zeros, or even mentally subtracting a zero from the printed scale, depending on the voltage range that the device is required to monitor.

This method of converting a meter designed to read current to indicate circuit voltage can also serve another purpose. Since the modification required to enable a meter to give voltage readings is done externally to the meter proper, the external resistor could be switched into the metering circuit when a voltage measurement is desired; at other times, the meter could be used to monitor the current drawn from the power supply. One meter can now serve two functions with the addition of a simple switch. In practical applications, one meter may be called on to indicate many different operating parameters, which are expressed in volts, kilovolts, amperes, milliamperes, and even microamperes. With the proper switching circuitry, one meter can adequately and accurately monitor all these values and give the proper indication when called on by switching in the part of the power-supply circuit from which an indication is desired. Through this method, a great savings in space is accomplished, as well as an appreciable savings in the cost of the circuit. Meters can be one of the most expensive single components of a small power supply. Meters can also be one of the largest usurpers of chassis or panel space. Multimetering circuits should be used wherever applicable. The simpler a circuit is made in actual construction, the lower the chance of a circuit breakdown.

Figure 6-3 is a schematic of a d'Arsonval meter used for voltage-measuring purposes. The small resistor indicated lies within the meter case and is the resistance of the moving coil, which is often referred to as the internal resistance of the meter. Figure 6-3 also shows the proper

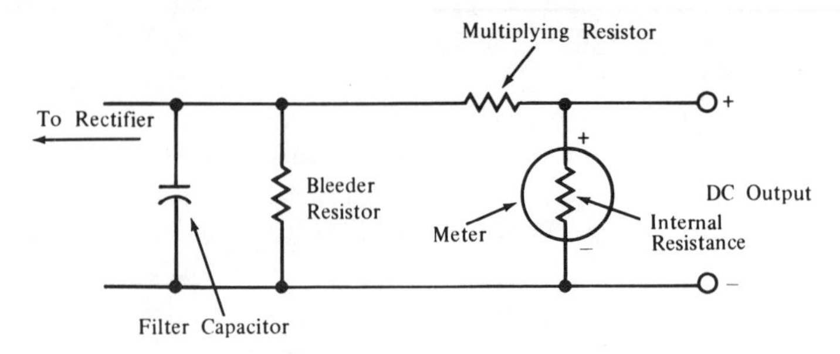

FIGURE 6-3. D'Arsonval meter set up to measure voltage potential.

place for the voltmeter when measuring voltage values from the output of a power supply. In this schematic, the voltmeter is connected directly across the positive and negative outputs of the power supply in parallel with the bleeder resistor, which is connected across the output in the same manner as the voltmeter.

INTERNAL METER RESISTANCE

The internal resistance of a meter is the resistance that lies within the moving coil. This resistance will vary from meter to meter and depends on construction and meter design sensitivity. Many times the internal meter resistance is ignored when calculating circuit values for metering purposes, and in noncritical situations the results will be adequate. Where accurate meter readings are a necessity, the internal resistance of the meter in question must be known. Manufacturers often include this information with the meter, but in many cases this is not done and some means must be found to determine the internal resistance of the meter to be used within a specialized circuit or circuits.

It is possible to measure the internal resistance of a meter with an ohmeter. Most ohmmeters, however, will, in the process of measuring resistance, pass a high current through the meter coil, which could cause permanent damage. Ohmmeters may be used for meters that read a relatively high value of milliamperes or amperes, but low-milliampere meters and microammeters will almost certainly be damaged. Another method must be found to give accurate readings of the internal resistance of meters without causing any damage to the meter coil or movement.

Figure 6-4 shows a method of determining internal resistance to a fairly accurate degree using a resistive bridge. The battery can be al-

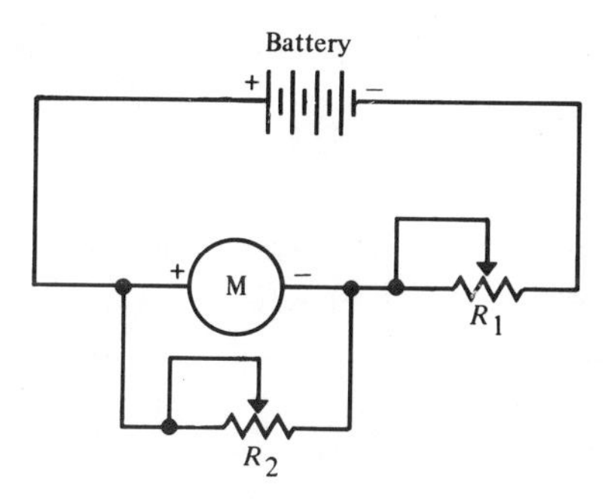

FIGURE 6-4. Resistive bridge used for measuring internal resistance of d'Arsonval meter.

most any small variety with about 1½ V of output. Connect variable resistor R_1 in series with the meter and set this control for a full-scale reading on the meter under test. When this has been accomplished, connect R_2 across the meter contacts and adjust this control until a reading of exactly one-half scale has been obtained. Carefully remove R_2 from the circuit, making certain that the control is at the same setting that accomplished the half-scale reading. Measure the resistance of R_2 with an accurate ohmmeter. The measured resistance will be equivalent to the internal resistance of the meter. A bridge of this simple nature cannot hope to compare with the commercial models available on today's market, but the internal resistance of the meter is obtained in an accurate enough form to assure very close meter readings in the circuit under construction or design. Control R_1 should be of the linear taper design and of a very high value of several hundred thousand ohms. Control R_2 should also be of the linear taper variety with a much lower resistance of 0–30 Ω for measuring the internal resistance of a microammeter. Values of 0–100 Ω work best on milliammeters. Generally, the resistance value of R_2 should be on the order of the anticipated internal resistance for the meter under test. Remember, do not connect R_2 into the circuit until a full-scale reading has been obtained by adjusting R_1.

CURRENT MEASUREMENTS

A voltmeter can usually be thought of as an external device that is placed *across* a positive and a negative point in the circuit to be measured. Voltmeters have a high total internal resistance and have little effect on the operation of the circuit because this circuit draws little current from the circuit being metered.

An ammeter can be thought of as an external device that is placed *within* the circuit and becomes a part of the circuit when installed. The internal resistance of an ammeter is very low in order to have as little effect as possible on the circuit. Applied current from a power supply will pass through the ammeter before it reaches the load. If the internal resistance were high, a large amount of power would be dissipated within the meter. If the internal resistance of a voltmeter were *low,* an excessive amount of current would flow, as it would act almost as a short circuit between the positive and negative sides of the power supply. A voltmeter has high internal resistance to have as little effect as possible on the power supply it meters. An ammeter has a very low internal resistance for the same purpose. Figure 6-5 shows one of several placement positions within a power supply for both a voltmeter and an ammeter. Both are actually current devices, with the external resistor in a voltmeter raising the internal resistance. This internal resistance has been raised so that it may be placed in the circuit in a manner that will allow it to measure voltage without having an adverse effect on the operation of the equipment being metered. In Figure 6-5, it can be seen that the ammeter is in series with the load and the voltmeter is in parallel or across the load.

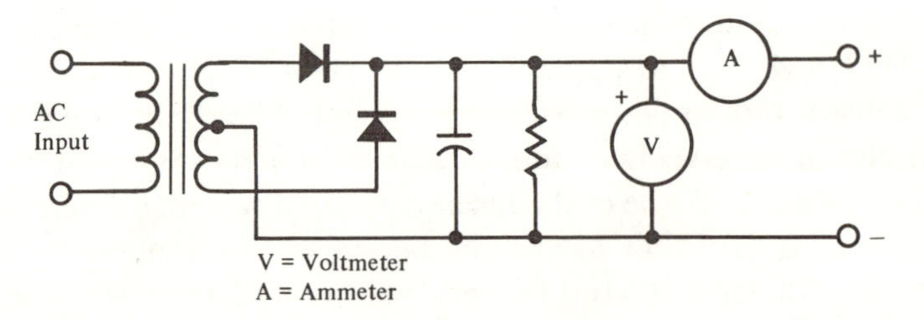

FIGURE 6-5. Meter placement points for voltage and current measurements in a power supply.

If a voltmeter is called on to measure more voltage than it was originally designed to read, the external resistor may be simply enlarged in ohmic value. A meter that was designed to read 100 V at full scale with a 100,000-Ω resistor can be altered to read 1,000 V full-scale by increasing the value of the resistor to 1,000,000 Ω; but an ammeter works a little differently regarding its connection within the circuit. It is in series with the load; therefore, adding a resistor in series with the unit is impractical, and subtracting the internal resistance is almost impossible by mechanical means. If a current of approximately 200 mA must be metered with a milliammeter that has a full-scale sensitivity of only 100 mA, the conversion is made with a shunt.

A *shunt* is a resistance that is placed in parallel with the moving coil in the meter. The shunt is quite easily connected across the two terminals of an ammeter, and is actually easier to install than an external resistor on a voltmeter. A shunt across a meter gives the circuit current two paths to take. One path is through the moving coil of the meter for an indication on the meter scale, and the other path lies through the shunt itself. For the problem presented above, the shunt must be designed with a resistance that will allow an even split of current, 100 mA passing through the shunt and the same amount passing through the meter when 200 mA of current is drawn from the circuit. The meter will then read full scale at a current drain of 200 mA, although only 100 mA actually travels through the moving coil. When a total of 100 mA is drawn from the entire circuit, half of that current will still be traveling through the meter and the other half through the shunt. The meter will read one half-scale, and so forth.

Figure 6-5 shows a typical shunt circuit across a milliammeter. The shunt is often a wound resistor, which depends on the inherent resistance of copper or steel wire to add ohmic losses to the circuit. If a reduction of current through the meter is required on the order of one half, as in the previous example, the shunt resistance will equal the internal meter resistance. A simple way of determining the shunt winding is to figure the internal resistance of the meter to be shunted by using the bridge circuit depicted in Figure 6-4, and then to figure the resistance needed for the shunt. Again, if a current division of one half is desired, the shunt resistance should equal the internal resistance of the meter. The formula used to determine the overall internal resistance seen by the circuit to be metered is the formula learned earlier for determining the combined resistance of resistors connected in parallel: $R_1R_2/R_1 + R_2$, where R_1 is the internal resistance and R_2 is the resistance of the shunt. Therefore, if the resistance of a milliammeter that has a full-scale reading at 100 mA is 100 Ω, to give a full-scale reading at 200 Ω, the shunt must also be 100 Ω for a combined circuit resistance of 50 Ω ($R_1R_2/R_1 + R_2$).

Furthermore, if the same milliammeter were to be used to measure current in a circuit that could approach values of 300 mA, the shunt circuit would have to be of a value of less ohms than would be exhibited by the internal resistance of the meter. The shunt would have to be wound so that, when 300 mA were being drawn from the circuit, only 100 mA would be allowed to pass through the meter coil; the other 200 mA would pass through the shunt winding.

By knowing the internal resistance of the meter, it can be seen that the shunt would have to be half the ohmic size of the internal moving coil resistance in order to pass twice the current. A proper ratio of internal coil current to shunt current must be established and then applied to the resistance of the shunt winding. Ohm's law will aid in the developing of shunt circuits by providing a check system to make certain that the correct amounts of current will flow in the different legs of the circuit. In the example given for the 300-mA circuit using the 100-mA meter, the shunt must pass twice the meter current, so the ratio of the shunt current to the meter current is 2 to 1. When figuring resistance, the inverse of this ratio is used, and the resistance ratio of the shunt winding to the internal moving coil is 1 to 2. If the moving coil resistance is 100 Ω, the shunt resistance value must be 50 Ω. Even the most complex ratio of current values can be easily applied to ohmic values when the internal resistance of an ammeter is known, and very high values of current may be measured with low-scale ammeters.

Shunts may also be switched in and out of ammeter circuits to allow metering over a vast range of current values. Multipurpose meters may serve as voltmeters during one operation, be switched to ammeter service for the next, and then to microammeter service at a later time depending on what the switching circuit does to the meter, in the circuit. Shunts enable the designer to save on available space and on expenses when assembling a complicated piece of equipment that requires strict monitoring of several different parameters.

A convenient copper wire table is provided in Appendix C to aid in winding shunts for metering circuits. Once the correct resistance value for the shunt has been determined, the chart may be referred to and the proper length of wire chosen to provide the resistance required for the winding. Shunts can be wound on almost any insulating form and mounted directly to the meter contacts or to a mechanical switch if switching arrangements are need to complete the desired circuit. Often a 1- or 2-W resistor of very high ohmic value is used as a mounting form. The turns of wire are wound around the carbon body of the resistor and soldered to the resistor leads at a point where they enter the resistor body. The wire leads are then connected to the meter terminals. The resistor will slightly change the overall resistive values of the shunt; but when resistor values on the order of one megohm or more are chosen, this variation is negligible for most applications. An extra length of shunt wire may be used to finely adjust the overall resistance of the shunt element for very exacting requirements when necessary. An accurate source of current is desirable for proper setup, with separate metering from this source an extra advantage.

A wire size must be chosen that will adequately handle the amount of current that will be drawn through the shunt windings. A very small wire size will exhibit higher ohmic values, but this small-sized conductor may be inadequate to pass the current required without excessive heating of the windings. If this is the case, a larger-sized conductor will be necessary, which will require more windings to obtain the same amount of resistance within the meter shunt.

Insulation is also a factor in meter shunts. Most shunts will be insulated with an enamel coating throughout the entire length of the wire. If closely spaced turns are used when winding the shunt, high-voltage applications may cause an arcing between these turns, which will result in an ineffective meter shunt. This applies only to voltage on the very high scale of about 3,000 V and higher. Close spacing may be effectively utilized in other applications where voltage values are held to a lower level. Any signs of heating or arcing should be immediately corrected to maintain proper operation of the shunt and its related meter circuitry. When resistor windings get hot, resistance value often changes, as does the atomic makeup of the copper molecules. A standard carbon resistor usually exhibits two types of resistance. One is a rating derived from the resistor operated at room or ambient temperature; the other rating is derived from heated operation. In most applications, this small change has little effect on the overall circuit; but in meter-shunting applications, a change of only a few ohms can cause inaccurate meter readings from the circuit being monitored.

The safety aspects of metering cannot be overemphasized. A meter, especially in a medium- to high-voltage power-supply circuit, normally carries these voltage values across its terminals. Care should be taken to assure that these contacts cannot be easily or accidentally reached by unknowledgeable persons. As discussed earlier, meter indicator needles are often hot, carrying the full voltage potential above circuit ground. Should the glass safety plate on the meter face become cracked or broken, immediate replacement of the broken part or the entire meter is necessary.

METER SWITCHING

A d'Arsonval meter can be used to measure both current and voltage to a high degree of accuracy. The main determining factor as to what meters measure is dependent upon where they are placed in the power-supply circuit. If placed across the dc output, voltage measurements may be taken if the meter is set up to read voltage, as was described earlier. When placed in series with the power-supply load, a meter will indicate current within the circuit. By using simple switching techniques, one meter can be made to function as a voltmeter, indicating the power-supply output voltage value, and as a current indicator, metering the output current to the load. To accomplish this, the metering points in the power supply must be brought to the switch contacts through conductors and channeled across or through the meter to be switched.

Figure 6-6 shows a simple switching circuit to accomplish these two metering functions. The switch is a double-pole double-throw rotary model that connects the milliammeter to different points in the power-supply circuitry. The meter shunt, R_1, is in series with the positive dc lead from the power supply; the voltage-multiplying resistor, R_2, is connected between the power supply's negative output lead and the meter switching point. The switch is shown in the voltage-indicating position. Notice that the shunt remains in the positive output lead at all times, but its resistance value is so low that it has no appreciable effect on the power-supply circuit. The shunt maintains a continuous circuit through to the power supply load at all times. It is possible to switch an

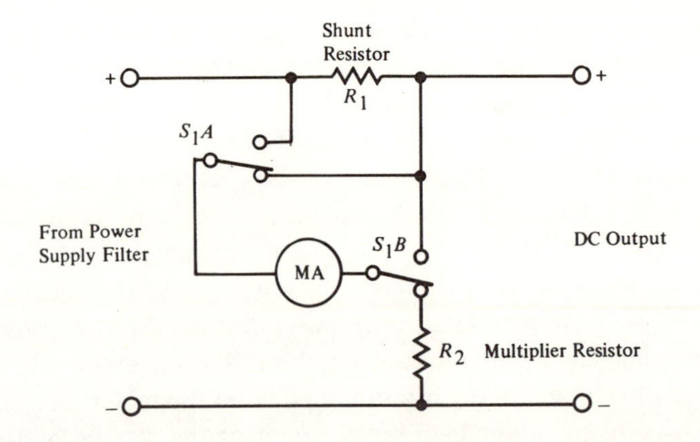

FIGURE 6-6. Meter switching circuit for multiple readings.

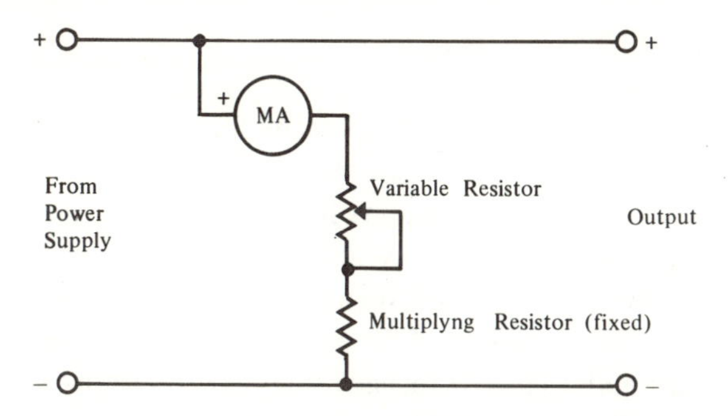

FIGURE 6-7. Meter calibration circuit used with d'Arsonval movement.

ammeter into and out of the circuit without using an external shunt; but the flow will be momentarily broken with each switching motion, and a special shorting contact setup would have to be included on the switch to maintain a continuous circuit while in the voltage-indicating position.

With the extra wiring required for a switched meter, often stray resistance is incorporated into the current path to the power-supply load and to the meter itself. This can cause variations in the voltage indication. This problem can be solved by installing a voltage-multiplying resistor that is slightly smaller than required by the formula. A variable control can then be inserted in the circuit between the voltage-multiplying resistor and the meter so that fine adjustment of the meter accuracy can be accomplished by slowly increasing the resistance value of the multiplying circuit. The resistor and the control, operating together, form the entire multiplying-resistor network. It is best to make this adjustment to the metering circuit as the final step upon completion of the power-supply and metering circuits. An external voltmeter of good accuracy is required. With this external meter placed across the power-supply output terminals, the variable control is adjusted to match the reading obtained. Figure 6-7 is a schematic of a simple adjustment circuit of the type described. The control should be of the linear taper variety and of fairly good quality. The resistance value of this control would be determined by the amount of resistance required from the voltage-multiplier formula. *Warning:* in high-voltage circuits, the variable control must be insulated to handle the voltage values and is a potential source of dangerous electrical shock should this insulation fail. Adjustment of the control with an insulated screwdriver is recommended to avoid this danger.

HIGH-VOLTAGE METERING

The cost factor of high-voltage meters is sometimes prohibitive when commercial units are used. A microammeter or milliammeter may be effectively used for high-voltage metering purposes by using external resistor networks, as was described for lower-voltage circuits. Figure 6-8 shows a much-used metering setup for power supplies of 2,500 V to around 5,000 V. A 0–500 microammeter is used for the meter indicator and a 10-MΩ resistor stack is used as the meter multiplier. One 10-MΩ resistor would be subject to voltage breakdown at this potential, so ten 1-MΩ resistors are wired in series to form a resistor stack that will adequately handle the high-voltage potentials. These resistors are usually mounted on an insulated circuit board for adequate arcing protection and safety purposes. This arrangement allows voltages up to 5,000 to be safely and accurately metered using a microammeter.

Another factor that affects meter accuracy when using multiplier resistors or resistor stacks in the circuit is the error factor or tolerance of the resistors used in the construction. Resistors are available that are accurate to 1 percent of their stated values. Many common resistors, however, have 20 percent tolerances, which can affect meter readings by a considerable amount, because their values may be 20 percent higher or lower than the stamped resistance value. Other common resistors may have a 10 percent tolerance factor. For metering purposes, resistors with a 5 percent tolerance or lower should be the only kind considered. These types are more expensive than those with higher error factors, but they are well worth the added expense when attempting to maintain a high degree of metering accuracy.

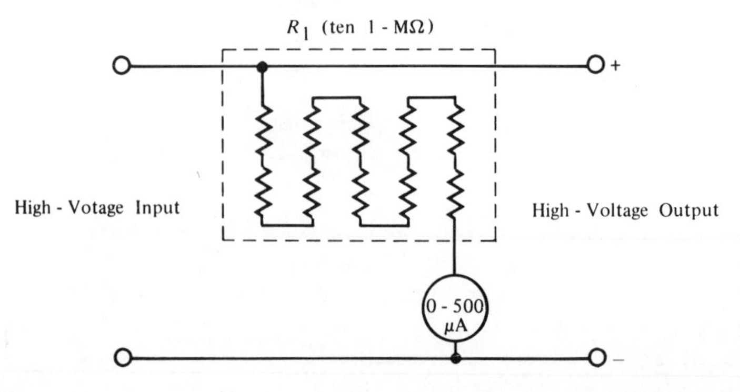

FIGURE 6-8. Typical metering circuit used with high-voltage power supplies.

METERING OF VARYING CURRENT SUPPLIES

In many power-supply applications, the current demand of the power-supply load will vary from a very low value to a high value in normal operation. Some meters are damped more than others and will be mechanically slower than others to respond to the current changes. When a device draws almost constant current from a power supply, meter response speed is not as important a consideration in choosing construction parts as it is when a varying current demand is required. Most modern meters of sound construction have adequate response to current changes for most applications; but where extreme accuracy and speed are requirements, a fast-acting meter should be considered. These meters cost a good deal more than their slower-acting counterparts, but are the only logical route to take when their use is dictated. In highly critical power supplies, metering is often a major factor in the cost of the overall unit. Do not neglect the quality factor in purchasing components for these types of circuits. Improper meter indications can lead to a cost factor in equipment replacement that may be many times the original cost of proper metering components.

Another factor to consider in metering circuits for power supplies that are called on to supply large amounts of peak current is the amount of current that is actually passed through the meter. Even the fastest-acting meter cannot read the instantaneous peak values of sudden current flows. They simply cannot act with enough speed. These peak demands may last for only a very small fraction of a second and may be equal to twice the amount indicated by the meter. Even though the absolute value is not shown by the meter indicator, this current is still passing through the meter coil. In peak current applications, an indication on the meter of 100 mA may mean that a value of 200 mA is actually passing through the meter coil for a brief instant. If the meter's range is only 125 mA, the meter coil is passing more current than it was made to withstand and may eventually fail. A meter with a top scale range of 250–300 mA would be a logical choice for this particular application. The anticipated current demands regarding peak values will be dictated by the type or types of equipment that the power supply is called on to feed. This factor must be taken into consideration when choosing the components for the supply *and* the metering circuitry.

Voltage values do not usually vary to the same extent that current does, and voltmeter circuitry does not require the same considerations. Any voltage fluctuations should be very minor in a well-designed supply, but even here a meter scale that is rated at a considerably higher value than the anticipated voltage to be read is very desirable. For instance, if a voltage of approximately 3,000 V or less is to be metered, a voltmeter with the capability of measurements to 3,500 V or more should be used. Variations in input line voltage could cause the dc output to rise to a value of 3,300 V or more in certain instances. A scale that is about 20 percent higher than the anticipated output voltage value of the power supply is a good rule of thumb when choosing voltmeters for power-supply measurements. This same percentage figure may be safely relied upon when choosing ammeters as well, but the 20 percent figure should be taken from the anticipated *peak* current that is to be delivered by the supply.

AUTOMATIC SHUTDOWN AND CONTROL THROUGH METERING

Modern meter design allows for many more features than simply monitoring power-supply operating parameters. Meters are now available with small light-sensing devices that acutate a switch when the needle indicator passes a certain preset point on the scale, which blocks a source of light supplied by a small bulb or light-emitting diode. When this occurs, a switch that is normally open closes its contacts, or one that is normally closed opens its contacts, to perform other functions. When this type of meter is used with an external relay, a circuit can be fabricated that will automatically shut off the primary power to the power-supply transformer and cease all power-supply operations. The meter setting can be adjusted so that when the power supply exceeds its normal current output, for example, operation is stopped before delicate electronic components in the supply or the load are damaged. Or the relay may activate a light or other warning device, which will alert an operator to a possible problem within the equipment. This type of meter may also be used to activate other equipment. For instance, an ac voltmeter of the switching variety may be set up to activate an alternative ac power source when the main source drops below a certain preset point. This could prevent overheating of motors and other voltage-dependent equipment.

The types of meters described usually cost several times more than conventional meters, but they also cut down on the price of external sensing equipment that may perform the same function. The requirements of specific operating equipment will dictate whether this is the most practical and least expensive route in building a power supply and metering circuit to perform the necessary functions.

DIGITAL READ-OUT METERING

More and more, the electronics industry is leaning toward digital read-out in many metering circuits. Digital read-out has advantages in many applications. It is generally more easily read than is the mechanical type of meter movement and is often more precise in actual measurements. Using compact, solid-state construction, digital meters often can take more abuse than conventional meters, and can be read in the dark without the necessity of external panel lamps when light-emitting diode displays are incorporated. Light-emitting diode (LED) displays usually display numbers that are constructed from several long sections of solid-state material that glow with a cool light when current is passed. The segments are arranged in a figure eight. Different segments of this display are electronically activated to produce a glow from the parts that will give the desired numeric indication. Figure 6-9 shows a typical digital read-out display chip with a pictorial explanation of how different numbers are formed. It can be seen that by using the figure eight as the central digit, all numbers from 0 to 9 can be formed and adequately displayed by lighting the different segments in the display chip.

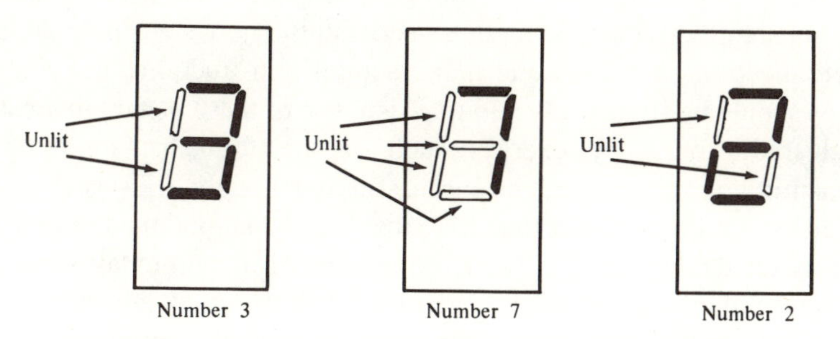

FIGURE 6-9. Display of numerals by a digital read-out chip.

The activation of the proper segments in the digital display is performed through an electronic circuit that utilizes integrated circuitry and acts on the input it receives from the portion of the power-supply circuit being monitored. When these meters are purchased as external power-supply components, it should be remembered that they are discrete electronic devices in themselves and require their own power source to supply current to the sensing devices within the meter.

Another form of digital display is the liquid-quartz read-out. Liquid quartz does not glow like the light-emitting diode display but reflects natural light from its surface. When current is passed through the solid-state chip, certain areas are sensitized and reflect light, creating a read-out comprised of black sections. These sections are usually arranged in a basic figure-eight pattern as described in Figure 6-9. Different areas of the chip are activated to reflect light from certain sections that form the proper number sequences, as with the LED display. A liquid-quartz display generally draws far less current than its LED counterpart and is popular for use with equipment that is operated from storage batteries.

Digital read-out metering displays are not generally used in applications that experience large value change throughout a normal operating cycle. In an application such as this, the digital display would continually count up to a high value and then count down again as the current demand increased and decreased. This often creates problems in getting an overall picture of what the power supply is doing; but for continuous values of voltage, current, and other operating parameters, the digital display is usually superior to the standard type of meter movement.

Electronic meters for the most part can be used just like meters of the moving-indicator variety. Voltage-multiplier resistors and shunt resistors are used with most models successfully; however, these are discrete electronic devices, and manufacturer's instructions and suggestions should be strictly adhered to. Also, use of these meters in areas of high radio-frequency energy, such as are found at high-power-transmitting sites, may not be successful owing to the RF susceptibility of electronic components that drive the display. Digital read-out meters are by far the highest-priced metering components on today's market; however, this added cost may well be worth the many advantages obtained when dealing with specific power-supply applications.

Application will also dictate the type of digital display used. If a continuous ac primary source is available, and a condition of dim lighting exists, the LED display would probably be the most desirable. If battery operation is desired and room lighting is not a problem, liquid-quartz displays would be more practical. Remember, liquid-quartz displays reflect room light and generate no illumination of their own. Many of these displays, however, are provided with an external light within the meter display case that may be activated when operating in areas of poor lighting. A small push-button switch is usually mounted on the meter case, which allows an operator to see the display when this button is depressed.

SUMMARY

Meters have many uses in modern power-supply design. Without proper metering, operation of power supplies and their electronic loads is impossible. Many transmitting applications require accurate metering of current and, in many instances, voltage to be in compliance with FCC regulations. Meters are usually rugged devices that require very little care and maintenance. Dust seems to be the only problem area with most conventional meters, but with proper cleaning and care, even inexpensive meters will withstand much wear and abuse and still continue to give accurate, reliable measurements of power-supply operating parameters.

QUESTIONS

1. List several reasons for the metering of power-supply parameters.
2. How is a magnetic field formed in a meter?
3. Most modern meters are are of one specific type. Name this type.
4. What factors determine meter sensitivity?
5. What purpose does the hairspring in the meter movement serve?
6. What operating position is most conducive to accurate meter operation?
7. All meters operate from what basic electrical unit?
8. What external device is often used with an ammeter to obtain voltage readings?
9. Meters should be chosen to give readings in what range of their total scales?
10. What purpose does a meter shunt serve?

11. What is the internal resistance of a meter?
12. How can internal resistance be safely measured?
13. How is a voltmeter normally placed in the power-supply circuit to obtain proper readings?
14. Where is an ammeter normally placed in a power-supply circuit to obtain its readings?
15. An ammeter with a high-scale reading of 500 mA is to be used to measure currents of up to 800 mA. To convert the meter to indicate 1,000 mA at high scale, what resistance should the meter shunt exhibit?
16. List the ways one meter may be used to measure voltage and current in the same circuit.
17. What physical conditions indicate a potentially dangerous condition at the meter regarding the safety of the operator?
18. What special considerations are normally given multiplier resistors when used in the metering of high-voltage circuits?
19. Do meters indicate the absolute value of peak current in varying load conditions? Why is this true?
20. How are certain meters able to perform a switching function when certain positions of the needle indicator are present?
21. What is the pivotal or base figure of a digital display?
22. Name the two popular types of digital displays. List the properties of both.
23. List some advantages of digital displays. Are there any disadvantages?
24. Why are meters possible danger areas where safety is concerned?
25. What is a moving coil?

chapter seven

SAFETY COMPONENTS AND CIRCUITRY

Throughout this text, constant stress has been placed on safety when building and servicing dc power supplies. Many of the higher dc voltages that are encountered in modern power supplies leave absolutely no room for error. Even the lower voltages can present a serious electrical shock hazard to inexperienced or careless operators. Although most mishaps with low to medium voltages do not often cause death, the potential is there, and given the proper set of conditions, even these voltages can be fatal. On the other hand, the high voltages present no possibility of a slight electrical shock. They are always an immediate threat to human life regardless of the duration of contact. Individuals who have been unfortunate enough to come in contact with voltages on the order of 1,500 V and more, and who have also been fortunate enough to survive a contact of this sort, do not merely have a healthy respect for high voltage, but a fear of its consequences.

Controlled fear is a much better attitude to have toward high voltages than healthy respect. In many truly sad cases, individuals who had a good working knowledge of electricity became careless owing to overconfidence. One mistake is all such individuals ever get. Electronics is a very safe hobby or occupation, but it can be totally unforgiving of even the slightest mistake.

Fortunately, many electrical and electronic devices and components are available on the market to help in the prevention of electrical mishaps with power supplies. Many of these are very simple items that can easily be constructed at the same time that the overall power supply is being built. When power supplies produce potentials of such a magnitude to place them in the high-voltage category, safety devices such as entry interlocks and automatic shorting bars to bleed off any excess charges of current lying temporarily dormant in the filter capacitors should be considered as mandatory safety equipment. Specific applications and uses may dictate many other safety features. It is generally considered good building practice to use electronic components such as transformers, rectifiers, capacitors, and resistors that are rated to withstand considerably more voltage or current than they will normally be called on to handle. Safety components should be given the same consideration. If two different types of door interlocks, for example, should be adequate to assure a safety factor, three interlocks will provide a safety tolerance should one of the other two fail.

The second most important safety factor for dc power supplies of any voltage is safety of the equipment, which includes the power supply proper and the electronic equipment being powered. If proper safety precautions have not been taken, electronic equipment can be severely damaged or, even worse, begin to burn. Electrical fires due to improper wiring and, more importantly, improper fusing, are reported in almost every city almost every day.

SAFETY INTERLOCKS

An *interlock* is a device that is mounted to the power-supply enclosure at a lid, door, or other route of access into the wiring, components, and voltage points. It is actually a switch that is triggered by the opening of the access lid or door to the power supply. Some interlocks are normally open while others are normally closed. When triggered, these devices reverse their normal states and either open or close the control contacts. A normally closed interlock is often placed in series with the plus or hot side of the filter capacitor chain and ground. As long as the voltage compartment accesses remain shut, the interlock stays in an open state; but should the access be opened, the contacts close, shorting the positive voltage lead with the negative lead and ground. The power supply immediately blows a fuse or trips a circuit breaker in the primary of the power transformer; at the same time, the filter capacitor string is being bled of all stored voltage through the direct short to ground.

This process assures that the line power supply to the transformer is no longer present, and the filter capacitor carries no charge if the technician has failed to deactivate the supply before entry. This should not be thought of as a replacement for the on–off switch or a shorting stick. This is an emergency safety precaution that puts a tremendous burden on the entire power supply each time it is triggered, with primary power being supplied to the transformer. Initially, the direct short circuit on the filter capacitor chain draws an extremely high amount of current through the rectifiers and the transformer before the fuse or circuit breaker has time to act. When an interlock system of this sort is triggered, often solid-state rectifiers or even stacks of rectifiers are permanently damaged to a point where replacement is necessary before operation can continue. Again, an interlock of this sort is triggered under power *only* when the operator or technician has carelessly failed to perform the proper safety actions before attempting to gain access to the power supply. The wear and tear on the power-supply components is regretful, but not nearly so much as the potential wear and tear on the human being careless enough to place himself in contact with a high-voltage point. When this type of interlock is triggered in a high-voltage circuit, the immediate array of sparks and cracking sounds is usually enough to prevent a recurrence in the near future of such a safety violation.

The other type of safety interlock, which has normally open switching contacts, is often used in conjunction with the type just described to break contact with the primary line voltage. The contacts are wired in series with one of the ac leads and the power transformer. When the power supply access is opened, this interlock opens its contacts and breaks the flow of alternating current to the supply, stopping all further operation. This interlock can be used as a secondary safety feature, but it does nothing to drain the filter capacitors of their stored energy. The bleeder resistors, if they are functioning properly, should perform this job; but if immediate contact upon opening the access door is made with the filter string, some charge may still be present that has not had time to bleed off, and a severe shock potential may still be present.

Combining the two types of interlocks provides a primary and secondary safety network that can protect the technician as well as the equipment. When the interlocks are tripped simultaneously, the primary ac voltage is disconnected at the same time that a short is placed across the filter capacitors. Damage can still occur to the rectifiers, but the overload factor on most of the power-supply components is lessened by a high degree. Figure 7-1 shows an example of interlocks in the primary voltage line and across the capacitors.

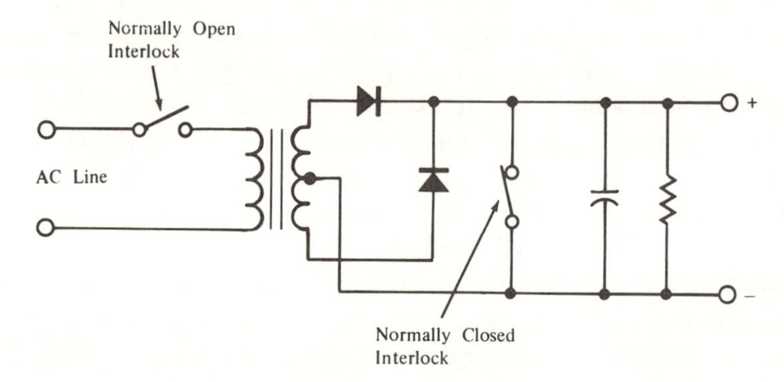

FIGURE 7-1. Power-supply circuit with safety interlocks installed at primary line and at filter capacitor.

Interlocks, as was previously stated, are simply switches that are actuated by an absence of the pressure that is supplied by a closed access door or lid. These specialized switches come in different sizes and ratings to handle different electrical requirements. A large interlock, one with a large surface area at the contact points, would be required for a high-voltage or high-current application. Smaller interlocks would be used for medium voltage–current equipment. Interlocks should be checked for proper operation and adjusted accordingly on a regular basis; this can be part of a regular preventive maintenance check on the entire power supply. A faulty electronic component within the power supply can cause equipment malfunction. A faulty safety interlock can cause technician malfunction. Make certain that all interlocks are working properly. An unprotected power supply operating at high voltages is potentially dangerous. A high-voltage supply that presents an appearance of having safety features is even more dangerous if those devices are not working, because some operators will feel that the device will protect them from any safety mistakes that might be made.

Never depend on an interlock or any safety feature to protect a human being from being killed or maimed. Interlocks have been known to fail, as have fuses, circuit breakers, shorting bars, and every other electrical part or component. When a power supply is shut down in the proper manner before service work begins, the interlock across the filter capacitors should automatically short any remaining charge to ground that the bleeder resistors have failed to discharge. If the bleeder resistors and the interlock should happen to fail at the same time, a lethal charge could still be present after the supply has been shut down. That charge could still be stored hours or days later. When service is required on a power supply, often it is because it is acting improperly or not at all. A component or associated wiring could be faulty. This problem could, conceivably, have caused the bleeder resistors to become defective or disconnected from their appropriate place in the circuit. A faulty interlock could be the final factor in a sad mishap if the technician does not follow proper service procedure by shorting the capacitors to ground with a heavy screwdriver or shorting bar. If a problem exists within the power supply, be highly suspicious of the safety features, interlocks, bleeder resistors and other components. Perform the required safety procedures *before* coming in contact with *any* voltage point.

The number of safety interlocks used with a power supply will depend on the number of entrances into the voltage compartment. It is a highly desirable feature to furnish two interlocks at each opening to provide a backup in case one should fail to operate. This is especially true if the power supply is to be used for training purposes.

AUTOMATIC SHORTING BARS

Automatic shorting bars are much like normally closed interlocks but are much simpler. Figure 7-2 shows an example of such a device. It is basically a piece of uninsulated aluminum that is bolted to the power-supply chassis in such a position as to make contact with the filter capacitor positive terminal or a point that connects to that terminal. When the power-supply access door or lid is closed, an insulator mounted to that cover makes contact with the shorting bar and pushes it away from the capacitor terminal, allowing normal power-supply operation. When the door is opened, the insulator is automatically pulled away, allowing the bar to come in contact with the terminal and shorting all the charged energy to ground.

The main problem with shorting bars of the handmade variety is

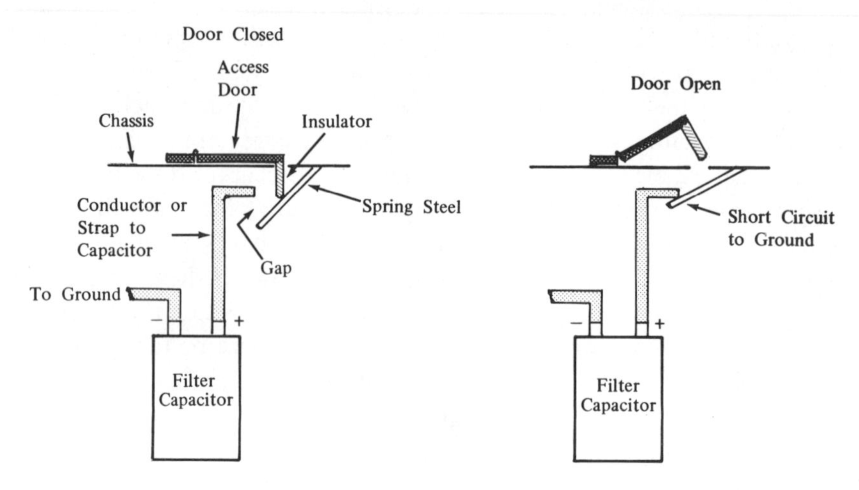

FIGURE 7-2. Automatic shorting bar.

their tendency to lose their tension or spring after repeated openings and closings of the access covers. Spring steel is often used and may be purchased at many electronic outlets. Spring steel is designed to maintain its tension even under severe or heavy-use conditions. Shorting bars of this type make excellent safety backups for access covers that are already protected by one or more interlocks.

Shorting bars as well as interlocks must be examined regularly for any signs of malfunction. A problem common to both types of safety devices is dirty contact areas. Oxidation may cause the contacts to become insulated from each other to a degree and hinder or halt proper operation. If these devices have been actuated under power-on conditions, the resultant arcing can pit and burn the contacts. Contact-cleaning compounds and burnishing tools should be used to clean these areas. The contacts should not only be clean but shiny to make certain that these devices will do their job if called upon.

Since most automatic shorting bars are constructed rather than purchased as manufactured components, a great degree of design flexibility is possible. The bar may be made in a size and form that will ideally suit the shape of the power-supply cabinet or housing compartment, and in a length to facilitate easy installation and access to the insulated separator that attaches to the power-supply cover. Care should be taken in choosing the insulator. Teflon makes an excellent high-voltage insulator, but mica- and ceramic-type insulators may also be used effectively. The shorting-bar arrangement is one of the simplest forms of safety devices available for power-supply usage and is quite easy to construct in a very short period of time. Owing to the large contact space generally available on an automatic shorting bar, it can be a more effective safety device than a normally closed interlock used for the same purpose. The difference is usually minimal, with the shorting bar having a slight advantage in less resistance to ground when the shorting process begins and, thus, a minutely faster bleed off of stored current. Again, the only problem that can present itself is a lessening of tension within the shorting section due to continued use. Service checks around the chassis connection on these bars should be made regularly to look for signs of metal fatigue, which will be indicated by a small crack or cracks near the mounting surface.

PRIMARY FUSING

When a short circuit occurs in power-supply circuitry, an increased drain of current is seen throughout, from the short back to the primary line input to the transformer. Most fusing of power supplies is done in the primary line. Fuses or circuit breakers do an adequate job of preventing actual burning of components due to a short circuit or constant current overload. It is nearly impossible to reliably test a fuse without actually blowing it, and it is often very impractical to test a circuit breaker without special equipment. Manufacturers' ratings are generally very reliable for both components, and they can be depended on to perform their jobs effectively if the correct rating is chosen for the power-supply circuit. A power supply that normally draws 2A of alternating current at the primary windings of the power transformer would probably be adequately protected by a 3-A fuse or circuit breaker; but if a 10-A fuse or breaker is used, the supply could conceivably draw slightly less than 10A with a serious short and rapidly heat the transformer windings or other components to a combustible point without blowing the fuse or tripping the circuit breaker.

Most ac lines are fused at the main service entrance, but this fusing should never be depended on to prevent damage to anything but the ac line running within the walls of a building. Proper ac primary protection is a must with all power supplies. This must be accomplished by using components of the correct value for the equipment being used within the power-supply circuitry.

Fuses or circuit breakers may be placed in either or both sides of a

115-V ac line to the power supply. When a 230-V feed is used with a third neutral or ground lead, the two hot leads should be fused; but never place a fuse or circuit breaker in the neutral lead. Even though the neutral fuse or circuit breaker blows or trips, a full ac potential of 230 V will still be present within the power supply if neither of the two hot leads has been opened. It is also very unwise to switch power on or off through the neutral feed.

The electrical ground for the ac supply should at all times be correctly connected to the power-supply chassis to prevent electrical shock. Make certain that the electrical ground is correctly established in the ac receptacle within the home or building that supplies this current.

Power cables that connect the power supply to the house current should be of adequate size and current-handling capabilities to withstand the amount of current that will be drawn. Cables that are too small will begin to heat, causing a fire hazard, and will probably deliver a decreased input voltage to the primary windings of the power transformer, with a resulting decrease in dc voltage output from the power supply. The copper wire table in Appendix C should aid in choosing cable with the correct size of conductor to be adequate for the current demands presented by specific power supplies where the ac input factor is known.

Another safety feature that is necessary is a main cutoff switch to control the ac line voltage feeding the power supplies. This switch should control all ac power at the outlets and be in plain view for easy access. Should a mishap occur, all power to the outlets may be immediately shut down by throwing this switch. This prevents continued contact with electrical circuits for the victim and prevents any further mishaps from occurring to individuals offering assistance. This master switch should also be plainly marked or labeled where possible. Some rewiring of present primary power feeds may be necessary, because in many instances two circuits may feed a single room in a home or building. Ideally, all power receptacles should be fed from a single voltage source that is controlled by the master switch. Figure 7-3 shows examples of good switching techniques for 115- and 230-V lines. When attempting installation of outlets, be certain to follow proper procedure, as specified by the electrical code or codes of your area.

The devices and methods just discussed assure that power will be turned off *before* entrance is made to the voltage points within a power

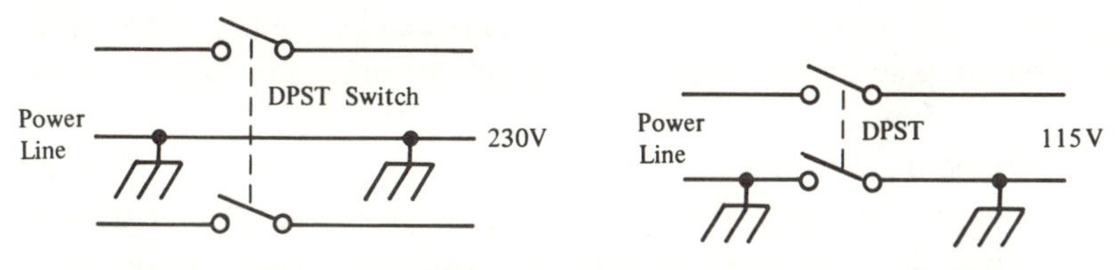

FIGURE 7-3. Proper switching techniques for 230- and 115-V lines.

supply. If any or all of these suggestions are followed, the possibility of an electrical accident becomes very low; but never depend on any type of interlock, relay, or other device to protect a human life. Make certain before coming in contact with any voltage point that no voltage is present. This can be done by using a shorting bar or even measuring that point with a voltmeter. Make certain that there is no doubt about the voltage potential at any point in the power supply when service is performed.

SERVICING ENERGIZED POWER SUPPLIES

Sometimes it is necessary to perform service work or adjustments on power supplies while they are energized. Certain adjustments require constant monitoring of output voltage or current, and, naturally, this requires that primary power to the power transformer be switched on and operation maintained until these adjustments are completed. The electronic load is another point where safety precautions must be practiced, because, even more than power supplies, electronic equipment that is powered by these supplies often must be energized to be adjusted. Whether the technician accidentally comes in contact with a voltage point from inside the power-supply compartment or from a power-supply lead inside the electronic load makes little difference. The potential danger is still coming from the power supply.

When performing service work on live power-supply circuits, proper tools and accessories will help to prevent mishaps; when these tools are coupled with caution and common sense, even high-voltage supplies may be adjusted in safety. Only high-grade screwdivers or probes should be considered for power-supply adjustment and service. Wooden handles should not be used. Wood does not provide adequate insulation at high voltages, and, should the handle become wet, could present a shock hazard at lower voltages. An insulated-shaft type of screwdriver is also an important item. This can be made from a typical metal-shafted screwdriver by winding tape or other insulation material around the shaft to a point about ½ in. from the tip. The screwdriver may now be used to reach to a point deep within the power-supply circuitry without fear of shorting any voltage points to ground with the screwdriver shaft. Make certain that the insulation material is adequate to protect from the voltages that will be encountered within the power supply. Several windings of tape may be necessary for higher voltages. Again, only handles of the plastic or rubber type are suitable for power-supply work.

Pliers, diagonal cutters, and other electrical and electronic tools should have handles that are insulated with plastic, rubber, or some other material that is rated to protect from electrical shock. Good tools may even carry a voltage-safety rating that specifies the insulation properties and the maximum voltage that these tools can be expected to be safely used with. The insulated-shaft screwdriver taping method may be used with other tools, including the ones just mentioned to prevent electrical short circuits, but a more intricate taping job will be necessary. Common sense should dictate which tools will or will not be adequate to maintain an acceptable margin of safety where power supplies are concerned.

A convenient item to keep in the tool box is a pair of rubber dishwashing gloves, which are sold in most stores. These gloves make excellent insulators when working around medium-voltage circuits where there is a *slight* possibility of a hand contacting an electrical point, such as a wire that may be old and losing a portion of its insulation. The gloves allow enough flexibility to manipulate small tools and provide an extra amount of protection to the operator or technician. Do not use these gloves around sharp contacts as they may easily penetrate through the glove to the operator's hand. Rubber gloves should be used as backup safety accessories only. They are not manufactured specifically for electrical purposes and are, therefore, not as reliable as other tools and accessories designed for working on electronic circuits and equipment.

A good working practice when servicing power supplies and other electronic equipment while energized is to keep one hand in your pocket at all times. The only way an electrical shock can occur is for a portion of the human body to come in contact with a positive point and a negative point at the same time. If one hand should rest on a positive point and the other on a negative point or point of potential difference, the electrical current will flow up one arm, through the chest, and down the other arm to the other contact. This channels the electrical current through the heart muscle. By keeping one hand in a pocket, the chances of receiving an electrical current through a large portion of the body are lessened. Any contact with points of potential difference within the power supply chassis would then be restricted to the fingers, hand, or arm. An electrical current that is confined to this area of the body is far less serious than a current that passes through the heart. Also, avoid the tendency to lean against the chassis. The ideal working situation is to see that no part of the human body comes in contact with any portion of the exposed power supply while energized. Less contact reduces the chances of an electrical shock of even a minor nature.

Make certain when cutting the power to the supply that all power is removed. Even the primary line voltage to the power transformer can be a serious hazard. Do not remove power by simply throwing the power supply on–off switch to the off position. Remove the power cord from the wall outlet or disconnect outlet power by throwing the main disconnect switch.

SUMMARY

Lack of knowledge seems to be the biggest cause of electrical accidents today. This does not mean that the persons involved in these accidents did not understand electrical and electronic fundamentals and practices. More often, they were not aware of a simple condition that existed at the time, such as the power supply being on instead of off, or an interlock being spliced out of the circuit, or a number of minor other events that were present at the time. Many accidents are caused by the ac line to the transformer primary. Individuals have thrown the power supply switch to the off position, assumed incorrectly that all power was absent from the power-supply compartment, and then come in contact with the primary voltage at the contacts of the on–off switch. This switch performed its function by interrupting power to the power-transformer primary, but alternating current was, of course, still present at the contacts on the power-line side of this switch. Many of these accidents seem to have occurred to individuals who made almost ridiculous errors in judgment, but further investigation has shown that many of these individuals were well qualified, accident free, and experienced. One or two factors could have contributed to their accidents. First, they may have been servicing equipment for long hours and become fatigued, sleepy, or even in a type of trance, which can occur from many hours of peering through small spaces, in poor lighting conditions, at tiny wires and components. Second, these individuals may have become overconfident in their abilities, disregarding proper servicing practices that they themselves would stress highly to the beginner in electronics. Any or all of these conditions can lead to a very serious accident. Safety precautions encompass the commonsense side of electronic building, experimentation, and servicing.

Do not continue to work after fatigue has set in. Do not service equipment when nervous, upset, or anything but sharp mentally. The trance-like condition previously mentioned is a state of mind that is familiar to most experienced technicians. It can be quickly overcome by taking a 15- or 20-minute respite from the work being done. When work is continued at a later time, the condition should not return. If it does, it is time to cease work for the day. Working under the condition described can lead to shaking hands and generally poor results. This condition probably is caused by a combination of fatigue, both mental and physical, eye strain, and muscle stiffness from remaining in one position for too long. There are many other outside factors that can affect working proficiency. When power supplies are being tested or serviced, any situation that affects work proficiency also affects the safety of the worker.

This chapter has stressed safety to a high degree. The safety factor in electronics cannot be overstated. Whole texts have been written on this subject alone. Electronic safety is easily learned, but it must be worked at constantly because its practice tends to be less heeded as actual electronic knowledge of circuits becomes more advanced.

QUESTIONS

1. Name some of the electronic protection devices in use today.
2. Describe the function of power-supply interlocks.
3. What maintenance should be performed regularly on interlocks?
4. What is a disadvantage of an automatic shorting bar?
5. Why is a malfunctioning safety device more of a hazard than no safety device at all?

6. What damages can occur when a safety interlock is tripped while the circuit is energized? What causes this damage?
7. What are the determining factors in choosing a primary-line fuse or circuit breaker?
8. What can be the consequences of a fused line that contains a fuse or circuit breaker much larger than is required?
9. How should a 230-V ac line using a three-wire system—two hot leads and a neutral—be fused?
10. What electrical and safety conditions result from using an ac line cord that is of insufficient size to carry the current drawn by the power supply?
11. What are the safety advantages of installing a main cutoff switch in the primary voltage line to power-supply outlets?
12. What should be a technician's main concern in choosing tools for service and adjustment work on dc power supplies?
13. Describe the advantages of an insulated screwdriver shaft when servicing power supplies.
14. Why do many technicians keep one hand away from the power supply while performing maintenance and adjustments with the other?
15. When servicing power supplies, can a technician safely work on a power supply that has been deenergized by throwing its on–off control to the off position? Why is this true?
16. List several factors regarding the technician's attitude and physical and mental state that contribute to electrical accidents.
17. What is technician's trance? What are its causes? What are its cures?

chapter eight

DC-TO-DC POWER SUPPLIES

Previous chapters in this text have dealt with different types of dc power supplies, but each has had one thing in common. Primary power for these supplies has always been a source of alternating current, standard 115- or 230-V house current in most examples. There is another type of dc power supply that has a pure dc output, as do the others, but which derives its primary power supply from another pure dc source. This type of power supply is often referred to as a dc-to-dc power source or power converter. The second half of these supplies is very similar to the dc supplies already studied, but the first half, the portion that accepts the dc input and transforms that voltage to a form that is usable with the conventional half of the circuit, is very different from every other basic power supply aspect already covered.

The dc-to-dc power supplies are used with electronic equipment that requires a pure dc voltage or voltages that are higher than the values usually found in conventional storage batteries. They are commercially available in many different voltage and current ratings for different applications. The dc-to-dc converters require more electronic parts than ac-driven supplies that deliver the same amounts of voltage and current, but many dc-to-dc supplies are smaller and lighter in weight than their ac counterparts.

Power transformers are ac devices that can increase the voltage value of their primary inputs, keep this value the same, or decrease it. This transformation can be accomplished only when an ac input or a voltage change is present at the transformer primary. The same is true of transformers that operate in dc-to-dc converters or power supplies. The reason the term *Converter* is used in describing dc-to-dc power supplies is due to the fact that the supply actually converts the direct current obtained from a battery or other dc source at the input to alternating current. Once this process has been accomplished, it is an easy matter to apply this generated alternating current to the primary of a transformer, rectify the output, and filter it, as is done with the other power supplies discussed in this text. The conversion of direct to alternating current will be a new subject to study in this chapter, as will the different component considerations that must be given even after the conversion from direct to alternating current has been accomplished.

If the reader continually remembers that once the current conversion has taken place, the procedures for rectification, filtering, and regulation are basically the same as in an ac-driven supply, understanding will come much more easily. Again, a dc-to-dc power supply has an input of pure direct current; this input is converted to a form of alternating current; this alternating current is transformed and converted back to direct current at a different voltage value than the original input. Two conversions have taken place in this type of supply rather than only one as in an ac-driven supply.

VOLTAGE INVERTERS

A device or component that changes direct current into alternating current is known as a *voltage inverter*. A type of inverter circuit is usually the first stage of a dc-to-dc converter. It changes the dc input to ac output for driving a transformer. Inverters may use a component called a *vibrator,* which is a small magnetic coil device that switches a thin metal strip back and forth between two other metal strips at a rate of 60 times per second or more. Figure 8-1 is a schematic of a vibrator. This device can be thought of as a fast-acting relay or switch of the single-pole double-throw variety.

The vibrator does not magically transform direct current to alternating current, but reverses the polarity of the dc input by constantly switching the positive and minus leads coming from the battery. Figure 8-2 shows how this is accomplished. A transformer with a center-

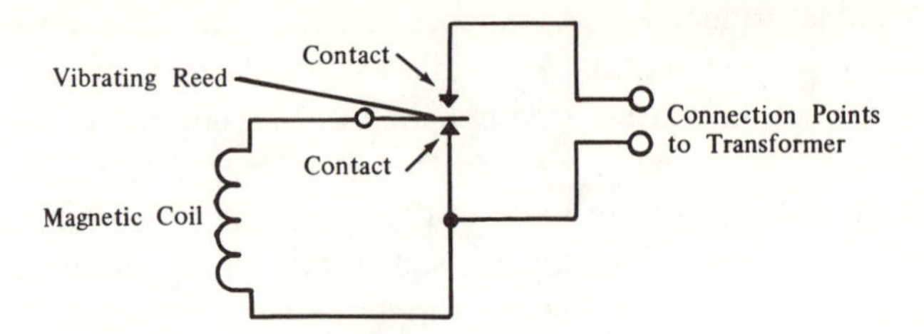

FIGURE 8-1. Simple vibrator.

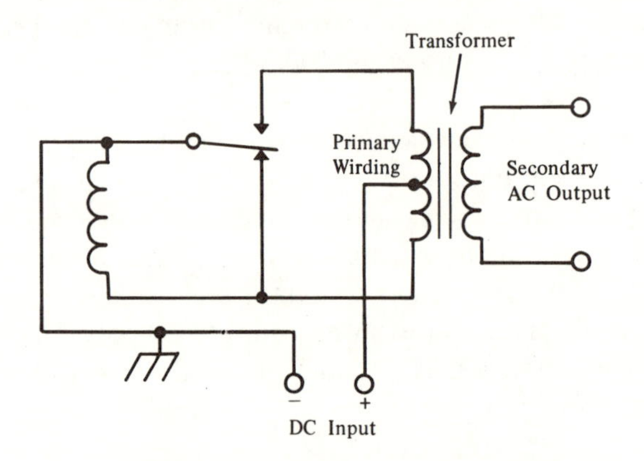

FIGURE 8-2. Vibrator inverter circuit.

tapped primary winding is used in conjunction with the vibrator. The center tap is always at a positive polarity in this example because the positive battery terminal is connected directly to the center tap. The negative terminal connects to the vibrating strip or reed and is switched back and forth between the two stationary contacts, which, in turn, are connected to the two sides of the center-tapped primary. In one position, the top transformer lead is negative in relation to the center tap. In the other position, the bottom lead is negative. This action produces a constantly changing magnetic field, and the transformer operates as if true alternating current was being applied to the primary.

Figure 8-3 shows the voltage wave produced by the vibrator and transformer and a true ac waveform. It can be seen that the vibrator output is not a true ac wave-form, but, rather, a square wave with a positive and minus side. A true ac sine wave builds up voltage peaks gradually, reaches a peak, and then decays gradually. This is the reason for the smooth curves in the waveform. The square wave, on

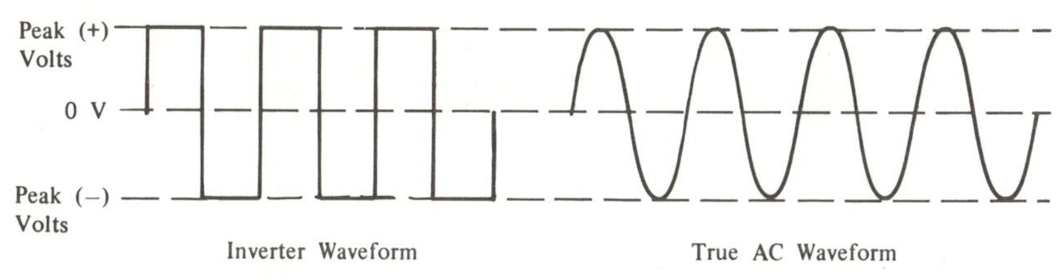

FIGURE 8-3. Waveform of inverter circuit compared with pure ac waveform.

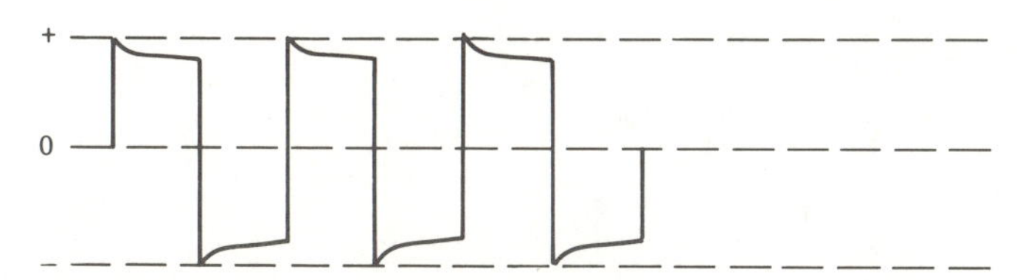

FIGURE 8-4. Actual inverter waveform under normal conditions.

the other hand, is either full on or full off. There is no gradual decay, and thus the sharp cutoff points. Technically, the square waveform that is seen at the input of the transformer in a dc-to-dc converter only has one polarity, plus or minus, as is true with direct current. Alternating current, as has been discussed in this text, rises to a peak positive point, decays to a zero value, and then progresses to a peak negative value before returning to zero again and repeating the process over. The square wave produced in a converter circuit is mechanically reversed in polarity by the reed contacts switching from one side of the transformer to the other. The wave that is seen as positive by the transformer for one half the cycle is the same wave that is seen as negative during the other half of the simulated cycle because it has been switched to the other side of the transformer. The ac transformer has been fooled into thinking that there is an ac input at its primary, because the magnetic field is constantly reversing, as in a true ac circuit.

In practical circuits in which, owing to resistance and other factors, the rise from zero to peak value of the square wave is not instantaneous nor is an instantaneous cutoff accomplished, some sloping effect of the curves is experienced. A truer drawing of the square wave produced in actual practice is shown in Figure 8-4. The polarity connection at the transformer center-tapped primary is the central polarity point, and may be operated at either a positive or negative value. The output from the transformer secondary will be nearly identical to the input regarding the shape of the waveform.

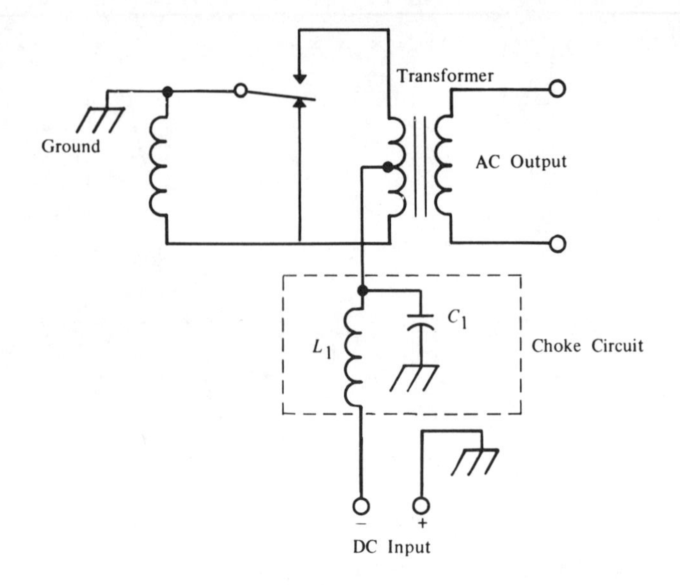

FIGURE 8-5. Inverter circuit showing hash filter.

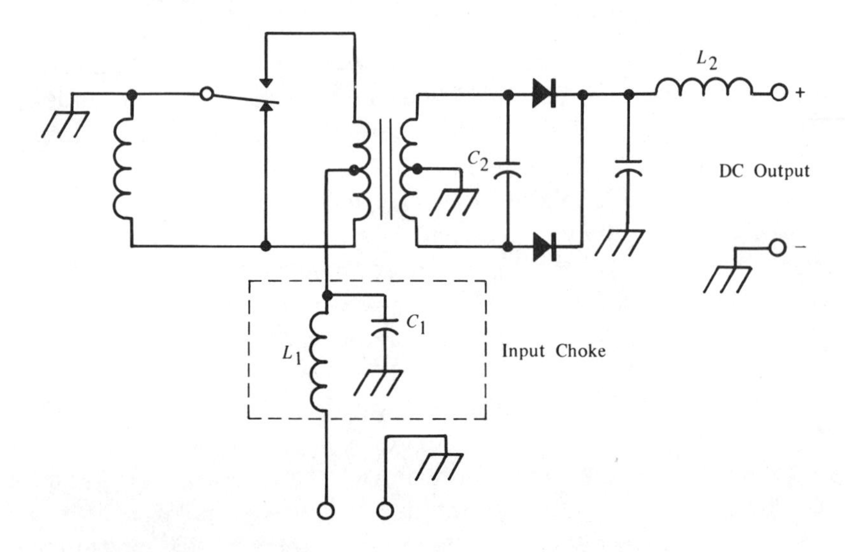

FIGURE 8-6. A dc-to-dc converter using input and output filtering.

The speed at which the vibrator makes and breaks its contacts is usually 60 hz, although many are designed to switch at a rate of 120 Hz. As with ac-powered circuits, faster switching speeds are more easily filtered after rectification.

Because of the mechanical switching of voltages across contacts, vibrators can create a great amount of noise or "hash," which can be heard in audio equipment as a popping or frying sound. This is due to the arcing across the vibrator contacts while the switching process is occurring. This arcing shortens the life of the vibrator contacts and renders this type of circuit useless for audio equipment unless proper steps are taken to bypass the contacts with a filtering circuit. A good filter will eliminate or greatly suppress these switching transients. Figure 8-5 shows a typical vibrator-switched inverter circuit with a filter made up of a capacitor and a filter choke. The choke suppresses the transients, while the capacitor shorts these unwanted currents to the circuit ground. For proper operation, this filter arrangement should be installed in an aluminum box and securely grounded. This forms a shield that will further suppress any noise generation.

Figure 8-6 carries the inverter one step further by adding a common full-wave center-tapped rectifier and simple filter arrangement to show how the inverter circuit is easily changed to a dc-to-dc converter, a circuit with a dc input and a dc output. Another filter choke and capacitor have been added to the secondary side of the power transformer to, again, prevent noise caused by arcing contacts. The capacitor across the secondary leads shorts the surges that occur when the magnetic field collapses upon breaking of the contacts. The magnetic field collapses so quickly that very high voltage surges occur for a fraction of a second. Without this capacitor, the vibrator contacts can actually be welded together by the heat produced in the sparking that occurs. This condition usually means that the vibrator is either damaged beyond reasonable repair or will require a laborious service job of separating the contacts, cleaning and burnishing them, and reinstalling the vibrator in the circuit. Many vibrators are of the sealed variety and cannot be repaired in a practical manner.

The type of vibrator that has been studied to this point is of the nonsynchronous variety. Another type of vibrator contains an extra set of stationary contacts that are connected to the secondary leads of the transformer. This synchronous vibrator eliminates the need for rectifiers, because the switching process that the rectifiers usually perform in an electronic fashion is accomplished in a mechanical manner by this extra set of contacts. Figure 8-7 shows a dc-to-dc circuit using the synchronous vibrator as the switching element. A similar center-tapped arrangement is used at the transformer secondary. This tap forms the positive output contact when the primary and secondary

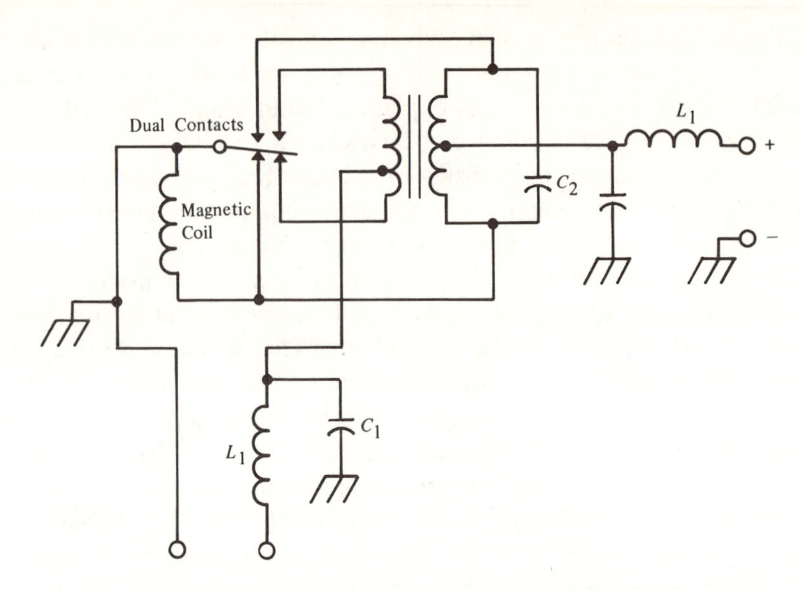

FIGURE 8-7. Synchronous dc-to-dc converter circuit.

polarities are synchronous. The proper connections for positive and negative leads are determined by experiment.

Filtering of switching transients is accomplished in a way similar to that of the nonsynchronous circuit discussed earlier. Voltage ratings of the second stationary set of contacts will determine the amount of voltage output that can be obtained with the individual vibrator, and will aid in choosing a power transformer that will adhere to these limitations.

Most vibrators have coils that are designed to operate from storage batteries with a voltage output of 6, 12, and sometimes 24 or 28 V. When activated, the coil pulls the center contact in one direction. When this first contact engages the stationary contact, energy to the coil is automatically short-circuited, the magnetic pull ceases, and the center contact or reed snaps back toward its original position; but the force of the snap causes the reed to travel past its central position, where it engages the other stationary contact. The coil energizes again, and the entire process is repeated. It can be seen that the storage battery voltage is actually used to drive two electronic components, the vibrator coil and the power transformer (while the switching process is taking place).

Most commercial vibrators are designed to deliver only a relatively small amount of current. The synchronous varieties usually operate in dc-to-dc circuits that deliver an output of between 100 and 400 V at output current ratings of up to 200 mA. Although this amount of current is on the low side for many load requirements, it should be understood that the primary contacts must withstand a current from the storage battery of about 20 A if the battery delivers a 6-V input to a converter that delivers 400 V at 200 mA at the output. Four hundred volts at a 200-mA current drain delivers 80 W of power to the load; but at 6 V, much more amperage must be drawn to produce the same wattage, plus the vibrator circuits are only about 65 percent efficient when comparing input power to output power. A 12-V vibrator would draw approximately half the amount of input current to deliver the same output; but even 10A is a very large amount of current for the relatively small vibrator contacts to pass without severe heating and voltage drop. The 80-W output from vibrator supplies is the upper limit for all practical uses. Higher current demands would necessitate the use of several different vibrator supplies operated with their outputs in a parallel configuration. The size and weight of such an arrangement would become prohibitive in most practical applications.

The voltage limits are fixed around the 400-V value because of the close spacing of the center reed to the stationary contacts. Higher voltages would present serious arcing problems within the vibrator assembly.

Vibrator-type dc-to-dc converters are a thing of the past. They are rarely used in any modern circuitry today. They were used heavily even into the early 1960s until they were replaced by smaller, more efficient transistorized converters. These modern dc-to-dc converters, however, work on the same switching principles as the vibrator supplies. This discussion on vibrator supplies will help pave the way for better understanding of the solid-state converters that will be studied in the next section. In their time, vibrators did an adequate job of providing the power needed for electronic equipment operated in a portable mode where the only source of power was the automobile storage battery.

Sometimes voltage-inverter circuits (direct to alternating current) are used as portable sources of alternating current to drive devices that normally are operated from standard house current. These inverters switch the dc input at a rate of approximately 60 Hz to more nearly imitate the conditions that are common to conventional ac sources. Power transformers are chosen for these circuits that operate on an input equivalent to the voltage value of an automobile storage battery, usually 12 V, and deliver an output at their secondaries of about 115 V. Many types of electronic equipment will operate from a portable source of this type, especially those which employ standard power supplies for transforming, rectifying, and filtering the original ac input.

Devices that contain motors often will not operate properly, because the motor speed may be determined by the frequency of the input voltage. This frequency is held to a very exacting value in ac house current feeds, but an inverter circuit of the vibrator type maintains only a coarse frequency tolerance. Some types of motors depend on a pure ac waveform and will not operate from the square wave presented by the inverter. Another reason for improper operation of motors is the amount of surge current that must be drawn to start their shafts in rotation. This large amount of current is needed for only a fraction of a second, but usually places too high a demand on the inverter circuit, causing component failure. The transistor circuits that will be discussed later are capable of higher output currents than most vibrator circuits, but even those efficient solid-state inverters usually cannot supply the current needed to start motor devices without being damaged.

When an inverter circuit is designed to deliver a voltage output and frequency that correspond to normal house current, power supplies that have been designed for ac inputs may be connected directly to the output of the inverter and will work almost as if they were connected to the house current outlet. This, then, provides a multipurpose use for conventional ac supplies. They may be used with standard ac sources or in a portable operation when connected to an external inverter circuit. This is basically what has been done in the vibrator-type converter circuits already shown, but the transformation of voltage was made in only one transformer instead of two in the example just discussed.

As is the case with vibrator-type converters, inverters using these components are rarely seen today. They too have been replaced by solid-state circuits that boast better efficiency of input-to-output power, quieter, more reliable operation, and compact size. A solid-state inverter capable of 100 W of output power at 115 V will often cost less than a comparable vibrator unit of the early 1960s, even in this day of high inflation.

SOLID-STATE INVERTER CIRCUITS

The same switching functions that are accomplished by mechanical actions and reactions in a vibrator may be accomplished electronically by using transistors and other solid-state devices to form an oscillating switch with no moving parts to reverse the dc voltage polarities to the primary taps of a special power transformer. Because of the solid-state switching, voltage arcing is eliminated and bypassing for noise reduction in audio equipment is a simpler task. Efficiency is improved because of reduced switching losses within the solid-state components.

A vibrator uses a magnetic coil to switch the reed back and forth among the stationary contacts; a solid-state switching circuit uses a feedback or oscillation transformer to set up an electronic means of switching on the forward-biased switching transistors. Figure 8-8 shows a schematic of a solid-state inverter circuit. The input voltage from the storage battery or other dc source is coupled to the emitter circuit of Q_1 and Q_2, the switching transistors. A voltage-divider network made up of resistors is used to set the transistors up in a forward-biased condition through the feedback winding on transformer

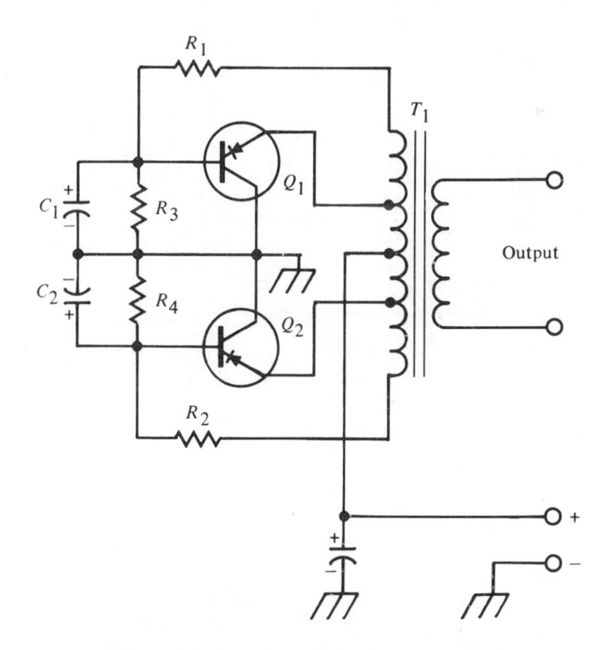

FIGURE 8-8. Solid-state inverter circuit.

T_1. Voltage is switched through the collector and emitter of transistors Q_1 and Q_2, which are connected in what is called a push–pull circuit. Each transistor conducts on alternate half-cycles of the oscillation. When one transistor is conducting, the other is in the off mode or blocking the flow of direct current. Voltage spikes are present in solid-state switching devices just as they are found in mechanical switching components.

Once the switching process has been started, the square-wave output is very similar to that obtained from a vibrator. This current is fed to the input of the power transformer T_2, where it may be stepped up to 115 V or any other value desired.

Many types of inverter transformers incorporate the feedback winding on the same core as the power transformer windings, making a

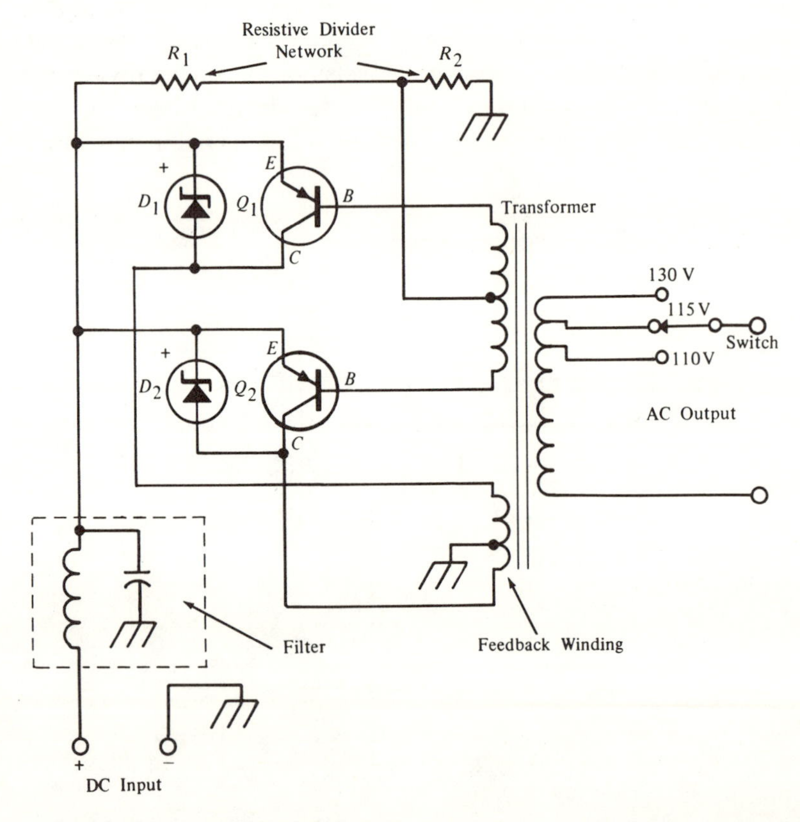

FIGURE 8-9. Solid-state inverter with multitapped secondary transformer winding.

more compact electrical component. The secondary windings of many of these transformers are also tapped at different voltage points to allow adjustment of the output voltage to a desired value. Low voltage at the input can cause the output voltage to drop below desired values, just as a higher than normal value of input voltage can cause the output voltage to be considerably above normal. These taps allow for switchable voltage adjustments in many commercial units, avoiding the necessity of removing the chassis from its cabinet and switching settings by soldering. Figure 8-9 shows an inverter circuit that uses only one transformer, with the feedback winding on the power transformer core and a switched-tap arrangement at the output.

Efficiency ratings of solid-state inverters can run upward of 90 percent. The amount of current drawn by the load will have a very large effect on this efficiency rating. In some circuits, ratings are high when only a small amount of current is drawn, as compared to the maximum amount that can be delivered by the supply. Other solid-state circuits are more efficient when near-maximum current is being drawn and much less efficient when smaller amounts are demanded. The frequency of the output voltage can also vary in the same manner as the efficiency ratings for different load resistances and, thus, current drains. The type of circuit used as well as the types and ratings of the inverter components will be a determining factor in the current, voltage, frequency, and efficiency ratings and fluctuations. Some circuit designs will provide excellent frequency stabilization and voltage regulation under load while operating at relatively poor efficiency levels; others will maintain a high average efficiency but offer only loose frequency stability. Circuits are usually designed to suit the individual needs of the electronic loads that they power; so a load that demanded strict adherence to frequency might concentrate on this factor while allowing the efficiency to be less than perfect. Or the opposite might be true for equipment that must be operated from marginal storage battery supplies that must last over a long period of operating time. Here, efficiency of input power to output power would be the paramount concern, and the inverter circuit would be built accordingly.

Many solid-state inverters are able to use very small power transformers owing to their fast switching rates. Circuits that are designed to deliver 60-Hz alternating current, as is present in standard house current, use larger transformers but are still compact and lightweight when compared to the outmoded vibrator supplies of the past. Small units that may be mounted beneath the dashboard of an automobile and occupy less space than a tape player are capable of power outputs of 150–200 W. Larger, more powerful inverters are used in pleasure boats, mobile homes, and sometimes private aircraft to supply electrical power for many electronic items that are normally used in the home.

As was the case with the vibrator types of dc-to-dc converters, solid-state inverter circuits form the first half of a solid-state converter circuit, with rectification and filtering of the ac output being the only remaining requirement for pure dc output.

DC-TO-DC POWER TRANSFORMERS

Before the discussion becomes heavily involved in actual converter circuits, a study of the specialized components that are used in these circuits is in order. The most specialized component in a dc-to-dc power supply is the power transformer, which is physically very different from the transformers discussed for the ac-powered supplies.

The main difference is often in the core material of the transformer, which may occupy a space several times smaller than a 60-Hz ac transformer. The frequency of the ac input current determines the size of the core and, often, the type of core material that can be used in practical construction. Generally, lower ac frequencies necessitate the inconvenience of a bulky core, which may be responsible for the great majority of the weight of a power supply. When the frequency of the alternating current is faster than 60 Hz, the size of the core can be effectively decreased for the same amount of power input and output. Faster switching rates also open the way for different, lightweight core materials to be used. The decreased size plus the lighter material makes very compact power supplies a reality where high-frequency alternating current or high switching rates are used. In a dc-to-dc converter, there is no need to use a 60-Hz switching rate, because the current from the transformer secondary will be rectified and filtered for a pure dc output. The frequency of the alternating current has no bearing on the rectified output current that is desired. The frequency does have a very large effect on converter design and component considerations, however, with the faster switching rates desirable in most applications.

When referring to standard ac-powered supplies, it has already been learned that a full-wave circuit is more easily filtered than a half-wave circuit, because the resulting dc pulsations from the rectifiers are at a higher frequency than those of a full-wave circuit. Half-wave rectifiers deliver pulsating direct current at a rate of 60 Hz in conventional circuits, whereas full-wave rectifiers produce 120-Hz pulsations. These pulsation rates apply when the input ac frequency is 60 Hz. In converter circuits, the input switching rate or frequency is often 400, 800, 1,000, or even 3,000 Hz and more.

Comparing a 1,000-Hz frequency with that of a common ac frequency of 60 Hz produces some interesting differences. At a rate of 1,000 Hz, one complete cycle occurs every 1/1,000 sec. One thousand hertz are completed, compared to a very slow 60 Hz for a standard ac rate in one second. With a 1,000-Hz switching rate, voltage applied to the primary of a dc-to-dc power transformer will produce a dc pulsation rate of 2,000 Hz at the unfiltered output of a full-wave rectifier circuit. The pulsation rate is much faster, and effective filtering is theoretically much easier and may be accomplished with much smaller values of capacitance than would normally be used for 60-Hz power supplies. Due to the nature of power supplies that use square-wave rectification, large values of output capacitance are recommended to avoid hum or whine, which is created by the ripple factor still present in the dc output; but regulation is accomplished to a high degree with smaller values.

Transformer core material is often made from powdered iron called ferrite. A ferrite core will often be formed in a circular mold so that the finished product looks like a donut. Turns are wound through the hole in the center and are spaced around the circumference of the circle. Figure 8-10 shows a typical core arrangement with the various turn windings. A ferrite core is used only for the higher switching rates of 400 Hz or more. The particles in the core are treated with a lacquer so

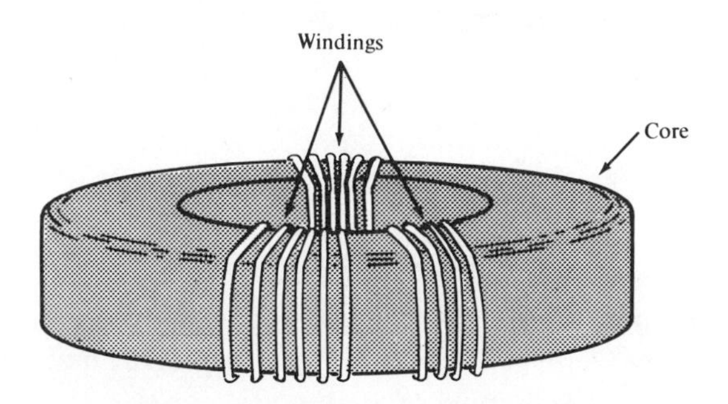

FIGURE 8-10. Toroidal core with windings.

that each tiny piece of iron is insulated from the thousands of others that compose the overall transformer core. This type of construction is less dense and lighter in weight than a solid iron core and gives excellent high-frequency response in power-supply circuits.

Windings are usually of the enamel-clad conductor variety as is used in the 60-Hz transformers discussed earlier. After the winding of the turns onto the form has been completed, the entire unit is encapsulated by wrapping with a high grade of electrical tape or by dipping the core in an epoxy sealing solution. This makes the core immune to many types of moisture and wet-weather conditions that would cause failure in many conventional power transformers. Since many transformers used in inverters and converters are used in outdoor and mobile operations, resistance to heat, cold, and, generally, adverse conditions is a prime construction factor. Weight and physical size are also important factors, and, fortunately, the nature of ferrite cores formed in a circular or toroidal configuration is suited to these requirements for portable operations. Toroidal cores are often manufactured in one general size and may be stacked, one atop the other, when a core of larger mass is required.

SEMICONDUCTOR DEVICES

Solid-state devices used in the switching circuits of inverter and converter circuits are generally of two types, transistors and silicon-controlled rectifiers (SCR). Some hybrid circuits use both devices to accomplish this switching. Circuits that use SCR devices usually work best at a voltage input of up to 700 V dc. Transistor circuits are generally used with voltage inputs of up to 175 V dc. Silicon-controlled-rectifier circuits may draw up to about 25 A of input current; transistor circuits often draw as much as 100 A on the input side. Both types of converters provide a square-wave output at the transformer secondary.

Output voltage from both types of switching circuits can be almost any value and depends on the turns ration of the transformer secondary. Output power can be anywhere from a fraction of a watt to 1,000 W or more. Output frequencies from the secondary of the power transformer will rarely be lower than 60 Hz, but may be as high as 100,000 Hz depending on the switching rate.

These nominal values may be increased considerably by wiring discrete circuits in series or parallel configurations. Theoretically, any amount of output voltage and current is possible, but the nominal figures given will encompass most power supplies in practical use today.

Some applications of inverters and converters will require an almost pure sine-wave output as is presented by a conventional ac power source. The ac output of these devices is usually a square wave, but elaborate filtering circuits can be incorporated in the units to arrive at a true sine-wave output. These circuits are used in many ways, but one of the most important is the control of frequency-sensitive motor circuits whose speed is controlled by the frequency of the ac sine wave. Square-wave operation causes these devices to operate erratically.

Transistor circuits are usually identified by the type of feedback circuits that they use to create the oscillation or switching process. Voltage-feedback circuits take a small portion of the output voltage and transfer it to the base junctions of the transistors for drive purposes. Current-feedback circuits sample a portion of the load current to drive the transistors at their bases. Current-feedback circuits normally use a separate feedback transformer and power transformer; voltage-feedback circuits can use a single transformer with all the secondary, primary, and feedback windings on a single core. Figure 8-11 shows examples of voltage- and current-feedback circuits in typical solid-state transistor inverters.

Due to the electronic advantages of both current- and voltage-feedback circuits, the latter is normally used for inverters because they are often called on to operate into inductive loads or devices containing transformers. Current-feedback inverters are more appropriately used in supplying power to resistive or slightly capacitive loads. Voltage feedback is most useful for electronic equipment loads. Most inverter circuits are load conscious. If operated under conditions that are not suitable for the particular inverter circuit in use, the oscillating action may fail to start, and one transistor may be damaged owing to the large amount of direct current being conducted. Other types of circuits may have built-in protection and just fail to operate without causing any component damage.

Again, conditions of operation will determine the proper type or types of circuit to be used in construction. Current-feedback inverters, for example, are more efficient than the voltage-feedback type for providing power to loads that present a constantly varying current drain. They are also better suited to conditions that experience changes in the

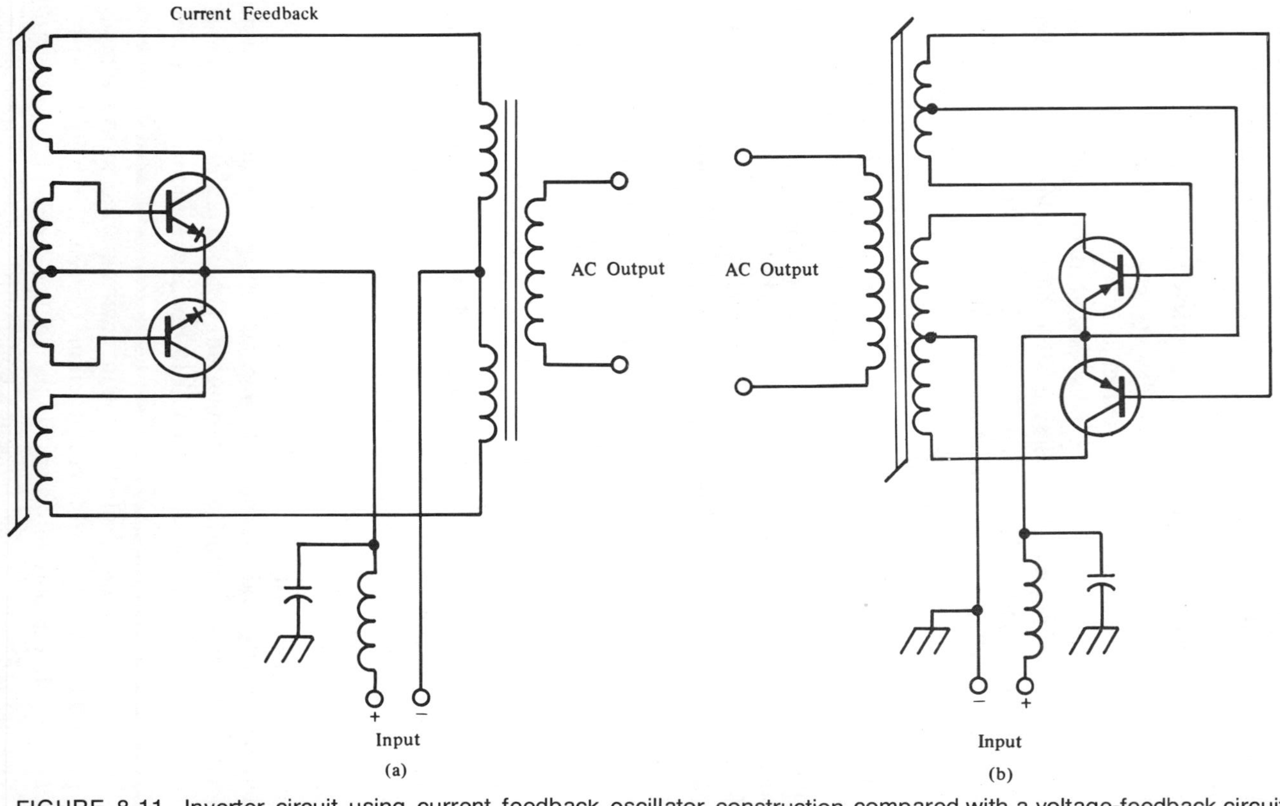

FIGURE 8-11. Inverter circuit using current feedback oscillator construction compared with a voltage-feedback circuit.

available input voltage to the inverter. Voltage-feedback inverters sample their base drive voltage from the output. When high current drains are in effect, the voltage has a tendency to drop owing to resistive losses in the conductors that make up the turns on the transformer core. The sampled voltage value also drops, and base drive may be inadequate. On the other hand, when current demand is zero or minimal, the output voltage will tend to soar. This condition is carried over to the sampled voltage and could possibly overdrive the switching transistor bases. Circuits that help to overcome these disadvantages are often used when, for other reasons, one type is preferred over another.

The convenience and lower cost of a one-transformer inverter circuit are difficult to maintain under conditions of high current output, where good frequency regulation and high output power are desired. For this reason, two-transformer inverters are often seen in high-power applications.

Silicon-controlled rectifier inverters are often used in operations that provide an input voltage that is higher than is practical for use with their transistorized counterparts. Basic circuits are relatively simple, with the most intricate subcircuits comprising the commutating functions that switch the activated SCR off. Once an SCR is made to conduct, it will continue this transfer or switching of current until the load current is decreased to a zero value. Series-commutated inverters use capacitors in series with the load to return the current value to zero. Parallel-commutated circuits use capacitors in parallel with the load to turn the device off.

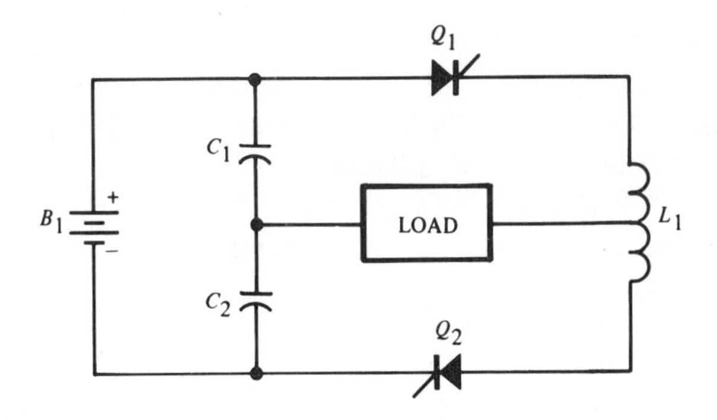

FIGURE 8-12. Series-commutated SCR inverter circuit.

Figure 8-12 shows a simple inverter circuit of the series-commutated variety. This is not a complete schematic; the drawing lacks the triggering circuit, which is often a simple multivibrator oscillating at the required switching rate to feed pulses to the gates of the SCR's to activate them into conduction. This schematic, then, depicts the main SCR switching devices, the ones that pass all the load current. Each capacitor C_1 and C_2 has half of the source voltage applied to its leads. When the top SCR receives its *on* pulse from the triggering circuit, half the supply voltage will be present across the upper half of the center-tapped inductor, L_1. C_1 will immediately start to discharge. As the top capacitor discharges into the load, the bottom capacitor, C_2, begins to charge through the load and the top portion of L_1. When the top capacitor has completely discharged, the bottom capacitor will be fully charged, and one half-cycle has been completed. The other half-cycle is a mirror image, electrically, of the first cycle, with the second SCR receiving its triggering pulse and discharging its stored energy into the load while the top capacitor is charged at the same rate through the load and the bottom half of the inductor. If the top capacitor discharges its energy in 1/120 sec, the bottom capacitor will do the same when it is discharging for a total of one complete cycle of positive and negative swing in 1/60 sec. The resultant ac output will have a frequency of 60 Hz. The load receives voltage on one half-cycle at one polarity, and during the other half-cycle at a reversed polarity, providing an ac input into the load.

This is a simplified explanation of the operation of a series-commutated inverter circuit. The charging values of each capacitor are not identical during the initial stages of operation, although the average voltage dropped by each capacitor will be one half the input voltage to the circuit. Although current distribution is not equal during the first stages of operation, peak voltage will continue to increase until each side of the cycle becomes equal. The output of a series-commutated inverter is a true sine wave, although it may not be as symmetrical as that obtained from a conventional alternating current source. The frequency of the output can be as high as 15,000 Hz before the SCR cutoff speed becomes a serious design hindrance.

Series-commutated inverters may take many different forms, but each of these circuits will follow the basic operating parameters and conditions as depicted by the example just discussed. Time-sharing circuits using series-commutated SCR devices are often used to provide an ac output. A sharing circuit usually describes SCR's wired in such a way as to have one current-carrying device operate on alternate cycles or alternate half-cycles while two others or two other sets of

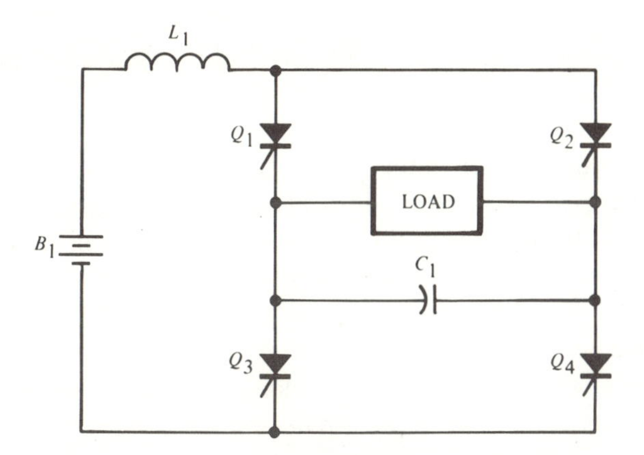

FIGURE 8-13. Parallel-commutated inverter circuit.

SCR's conduct during the remaining cycles. This can increase frequency output to the load and create a cooler operation for all components involved in such a circuit.

Figure 8-13 shows a representative parallel-commutated inverter circuit minus the driving network for turning the devices on. Only one capacitor is used in the circuit, which is connected in a parallel configuration with the inverter load. Four SCR devices make up this circuit, and they are operated in pairs that are triggered on alternate half-cycles. Q_1 and Q_4 form one pair, and the remaining devices make up the other. The inductor L_1 is not the center-tapped variety, which was used for the series-commutated circuit, and receives the full voltage potential of the input, which is in series with the remainder of the total circuit. When Q_1 and Q_4 are turned on by an external pulsing circuit, current flows through the inductor and progresses on through Q_1, the capacitor, the load, and finally through Q_4. When the remaining pair of SCR's is turned on, the capacitor begins to discharge through Q_1 and Q_2 and also through Q_3 and Q_4. Reverse current will then flow through Q_1 and Q_2 until the capacitor completely discharges and these devices revert to the off condition. At this point in the cycle, current will flow through Q_2, the capacitor, and Q_3. The current is still in the former or original state and will remain so until the current flowing through Q_2 and Q_3 charges the capacitor in the reverse direction. The first cycle is completed and operation begins again at the established switching rate that is set up by the external pulsing circuit and the capacitor across the load.

This fairly complicated circuit can be made simpler if the load is in the form of a center-tapped transformer. In this case, two of the SCR's can be eliminated, as the return from the transformer primary windings will complete the circuit.

SCR inverters are far outnumbered by their transistorized equivalents and are used mostly for industrial applications. Transistorized inverters are more relevant to this text, because they are, generally, of simpler electronic design and dc output when used in a converter circuit. There is little difference between dc output that is rectified from a square-wave ac output and one derived from that of a pure sinewave.

To summarize this portion of the chapter on converter circuits, dc-to-dc converters are powered by a pure source of direct current that is switched in some form, solid state or vibrator, to produce an alternating current in an internal inverter circuit. The alternating wave is usually in the form of a square wave, but may with special circuits be a true sine wave. Switching rates of inverters vary from 60 or 120 Hz, with the outmoded vibrator switching devices having up to many thousands of cycles per second when using modern solid-state circuitry. The higher switching rates with most of these devices allow a different core material to be used for transformers and other inductive devices such as chokes. Filter networks are often necessary when switching rates fall within the audio spectrum (300–15,000 Hz) to prevent these frequencies from carrying over into audio-sensing electronic loads. Solid-state inverters are usually very efficient devices that deliver upward of 90 percent of the input power to the load, as compared to about 60 percent for vibrator-switched circuits.

The understanding of the basic operating principles of inverter circuits is necessary before moving on to dc-to-dc converters. Study and learn the information provided before continuing on to the next portion of this chapter.

DC-TO-DC CONVERTER CIRCUITS

A dc-to-dc converter simply uses all the information discussed so far on inverters, rectifies the ac output, filters it by conventional means, and delivers the pure dc output to the load. Figure 8-14 shows a schematic of a finished dc-to-dc converter using a single toroidal transformer and switching transistors with their operating circuits. The output from the transformer is rectified through a standard full-wave bridge circuit, where filtering is accomplished by a single capacitor. The bleeder resistor is the final component before current is delivered to the load. This circuit uses voltage feedback to start the oscillation process for current switching within the transistors.

After voltage inversion has taken place, the ac output may be treated in the same manner as is done in ac-driven power supplies. All the rectifier combinations discussed previously are perfectly applicable to converter circuits. If the transformer does not provide a high enough voltage output for a specific application, voltage-multiplier circuits (preferably of the full-wave type) may be used to provide the correct values. All component ratings apply in a like manner, as do the safety

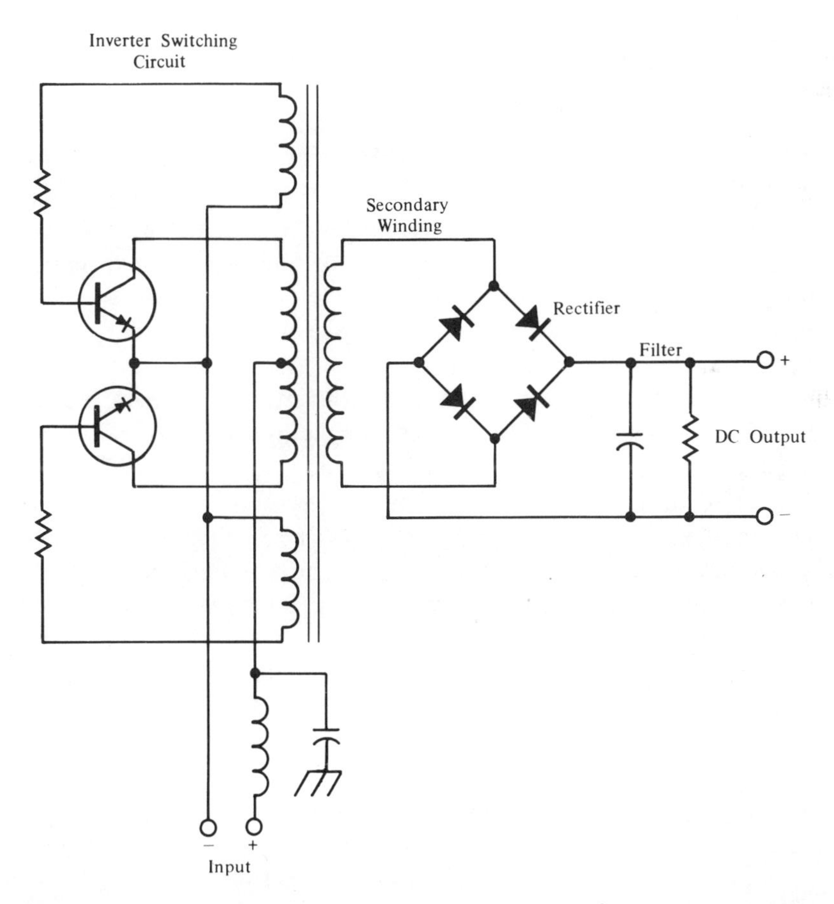

FIGURE 8-14. Complete dc-to-dc converter with full-wave bridge rectifier output.

precautions for working on live circuits. By referring to the schematic, it can be seen that the circuitry used immediately after the power transformer is identical in every way to that of conventional power-supply circuits and operates in a like electronic manner, performing the same basic functions of rectification, filtering, and regulation.

The dc-to-dc power supplies must be thought of as being more delicate than ac-driven supplies regarding load demands. In an ac supply, a sudden current drain due to a short circuit or other malfunction will probably only trip a circuit breaker or blow a fuse. At the very worst, in a properly protected supply a rectifier may be damaged. In a dc-to-dc supply using solid-state devices, a like surge could damage the switching components before a protective fuse has time to open the circuit. Most solid-state devices are very rugged when operated at proper levels and last many times longer than tube-type equivalents, but, unlike vacuum tubes, solid-state devices are very vulnerable to sudden overloads in operating current. It is often said that transistors make excellent fuses because they often fail before fuses do, thus protecting the circuit from further damage during periods of overload. This statement is certainly true, but it should also be remembered that transistors and other solid-state components cost much more than fuses or other protective devices. When a protective device is correctly placed in solid-state circuits, it protects the equipment from a possible fire, but does little to protect the solid-state components.

Figure 8-15 shows another dc-to-dc converter circuit using a power transformer that has several secondary windings for delivering different output voltages. Each winding is rated at a specific voltage and maximum current value. Many such arrangements are available commercially that will supply a high voltage of between 600 and 800 V dc, a medium voltage of 250–350 V, and a control voltage of about 50 V dc. This type of supply is used most often to supply current for tube-type communications equipment.

The output power that can be drawn from dc-to-dc converters is, of course, dependent upon the component ratings, but the biggest limitation to high output is often the available dc input voltage. Switching transistors may be rated at a current of up to 100 A, with voltage ratings to 100 V or more, but most storage batteries deliver an output voltage of 6, 12, or 24 V dc. The current rating must be adhered to regardless of the voltage value at the input to the supply; therefore, in a circuit where transistors rated at 10 A are used, an input voltage from a storage battery of 12 V will allow a maximum power input of 120 W

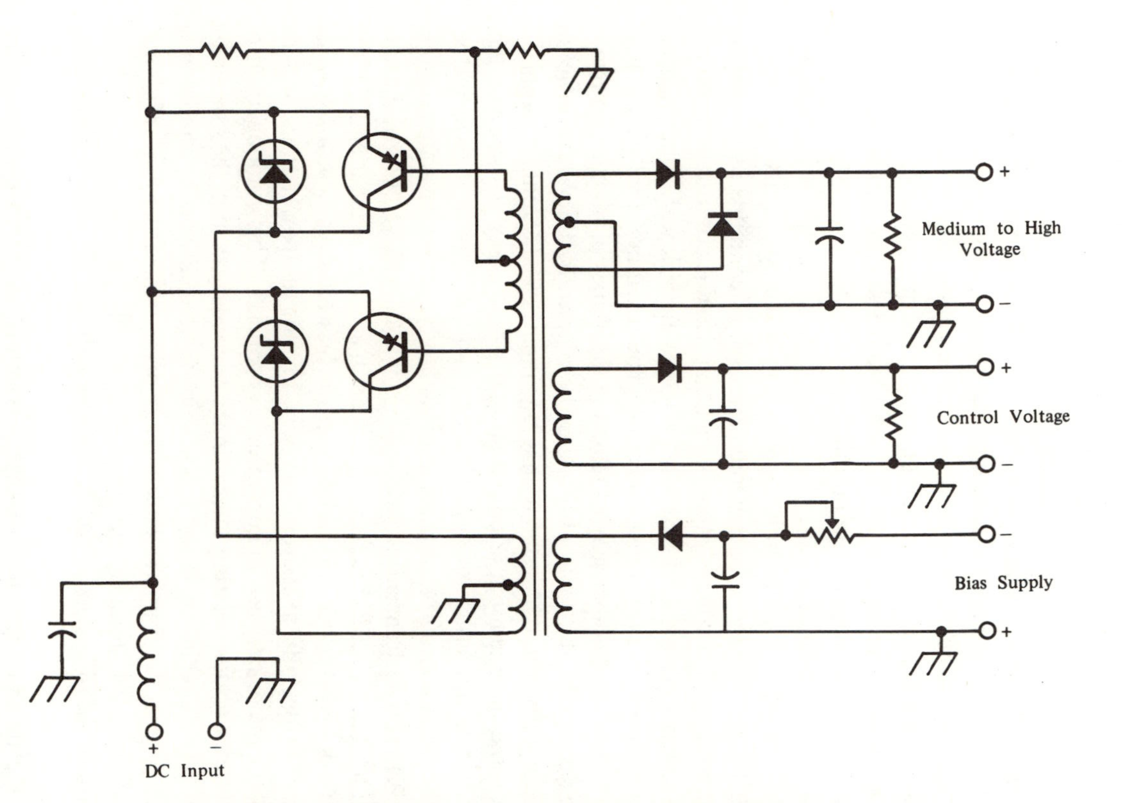

Figure 8-15. Converter circuit with several secondary windings for multioutput.

$(P = IE)$. Assuming a 90 percent efficiency factor of the converter circuit, the maximum power that could be delivered to the load would be 120×0.90, or 108 W. Using the same transistors with an input voltage of 24 V dc, the maximum input power to the converter would be 240 W, with an expected output of 216 W to the load. The power transformer would have to be changed in order to maintain the same voltage output as with the lower voltage input, and the components remaining would have to be rated to withstand the higher current drawn from the supply; but, basically, the power output has been doubled while using the same transistors. This assumes that the transistors are rated to operate at voltages as high as 24 V. Most transistors are limited only by their current ratings in these circuits.

When designing a dc-to-dc converter, problems that develop with high-power applications may vanish if the possibility of using a higher input voltage is considered. The cost of components to operate in a circuit with a voltage input higher than 12 V is not usually appreciably more than the cost of those designed for a lower voltage. This applies to the power transformer that would require a 24-V primary and little else. Another way of realizing the values of higher input voltages is to take the same example as before, with 108 W available at the output of the converter with 12 V input. When going to a 24-V input, the same amount of output power, 108 W, may be drawn by the load while operating the transistors at half the current it took in the former circuit to deliver the same amount of power at the output of the converter. The transistors, one of the most expensive components in the converter circuit, are being operated in a cooler manner and should give a longer service life. Also, there is less resistive heat loss in the circuit, especially in the primary windings of the power transformer, another major expense in converter components. This component also operates with less heating effects in the secondary than in the other circuit. Whenever there is an opportunity to reduce heat in almost any electronic circuit, more reliable, steady performance and longer life are usually the long-term results.

Many portable sources of dc voltage will not allow the higher voltages to be used, but even in automobiles, which generally use 12-V storage batteries for their electrical systems, an increase in available voltage may be had by installing another battery in series with the one already present within the automobile, offering a 24-V dc source for portable operation of electronic equipment of a power rating that would make a higher input more practical for design considerations. Some converters are designed to operate from a series bank of storage batteries, which may supply an input voltage in excess of 100 V dc. These systems offer unusually high power levels at the output of a single converter or even several converters operated with their outputs in parallel for a multiplied power output.

SUMMARY

For the most part, it can be seen that dc-to-dc converters are simply dc-to-ac inverters with the voltage output rectified and filtered by very common methods. The solid-state converter, however, is one of the most important improvements to power supplies since the invention of solid-state components. More and more, any components that contain physically moving parts, such as the vibrator, are being replaced by devices that move current through solid-state switching. The biggest detriment to even smaller dc-to-dc converter packaging is heat. Many times the largest physical part of a converter circuit, especially those that draw large amounts of input current, is the heat sink(s) for the switching transistors. These may easily be as large as the remaining portion of the circuit, but without them the switching transistors would be destroyed in a very short operational time. Still, dc-to-dc converters are smaller and less bulky than portable power sources of years back. During that time, a power supply, whether of the ac or dc input variety was usually the largest and heaviest component in any electronics system. Today, it can be one of the smallest.

QUESTIONS

1. Define a voltage inverter.
2. How does an inverter perform its function?
3. Name the three devices or components that can be used to perform voltage switching processes in inverter circuits.
4. Generally, how does a vibrator work?
5. What type of output waveform does a vibrator-driven inverter have?
6. What type of primary winding must a power transformer used with a vibrator-driven supply have?

7. What speeds do vibrators normally switch current at?
8. What are the disadvantages of using a vibrator type of inverter circuit as compared to a solid-state unit?
9. What is the difference between a synchronous and nonsynchronous vibrator?
10. What is the approximate input-to-output efficiency of a vibrator-type supply? *Problem:* A vibrator-type inverter draws 15 A of current at 6-V input. Assuming a normal efficiency for this circuit, what will be its approximate power output?
11. What types of devices are usually not operable from inverter circuits?
12. List some advantages of solid-state switching in inverter circuits.
13. What efficiency ratings do solid-state switching circuits in inverters often exhibit?
14. Why do many solid-state inverter circuits use small toroidal transformers?
15. What core material is used in toroidal transformers?
16. What values of input voltage and current would be best suited to an SCR-switched inverter circuit? to a transistorized circuit?
17. What waveform is usually present at the output of a transistorized inverter circuit? of a series-commutated SCR circuit?
18. List the two types of transistor switching circuits.
19. Where is a two-transformer transistor-switched inverter circuit most applicable?
20. What types of rectifier and filter circuits may be used at the output of an inverter circuit to form a dc-to-dc converter?
21. How subject are dc-to-dc converters to current surge damage?
22. A dc-to-dc converter has an input voltage supply of 12 V. The transistors in the switching network have a voltage rating of 50 V and a current rating of 10 A. What is the maximum power that may be safely handled by these transistors?
23. List the advantages of using as high an input voltage as possible with dc-to-dc converters.
24. Assuming an efficiency factor of 90 percent, what is the amount of power that can be delivered to the load in a converter circuit that draws 20 A at 32 V from the input power source?

25. The outputs of two identical converters are wired in parallel. What will the total combined output voltage and current rating be if a single unit is rated at 400 V and 400-mA output?
26. If the secondary of an inverter circuit delivers 225 V at a current rating of 550 mA, what will be the approximate value of voltage and the current rating after this ac output from the transformer has been rectified by a voltage-tripler circuit and filtered under conditions of no load?
27. Why are vibrator-driven inverters and converters outmoded?

part two

POWER-SUPPLY CIRCUITS

Basic theory and application techniques have been discussed in detail up to this point in the text. This information is necessary to know before moving into the next phase of study on power supplies designed with specific components for specific purposes.

The next three chapters will include much of what has been learned so far and will discuss actual power-supply circuits from the very basic to the relatively complex. Although the circuits included are designed to provide a more explicit schematic example of the theory discussed earlier, any of these power supplies may be built and operated as represented. These are schematic diagrams of actual working power supplies. A parts list is provided that gives the values of components not listed directly on the schematic, as well as any other pertinent information.

Each power supply will be described as to its particular operation in a step-by-step process that will start at the input and end at the output, while thoroughly covering each electronic step in between. By building one or more of the power supplies in each chapter, a better understanding may result of this entire text. It is much easier to apply knowledge to a working model or example. Although each schematic is of a power supply that has been tested for operation, minor changes in circuitry may be necessary owing to variations in different types of manufactured components and the like.

This text has dealt with the theoretical operation and possible applications for power-supply circuitry. For actual construction, sound building practices are necessary in order to end up with a properly working, safe power supply. If the builder is unfamiliar with electronic building techniques, a text on the subject is recommended along with assistance from an experienced builder or technician.

While building any of these power supplies, learn as construction takes place. Follow the logical current paths. Visualize the schematic and theoretical representation of the physical components and circuits being constructed. Follow the logical steps of transformation, rectification, filtering, and regulation as these portions of the supply are completed. When you can place yourself within the power-supply circuitry like one of the millions of electrons that flow to create electrical current and visualize the path that you will take throughout this circuitry and the changes that will be made before leaving the power supply to the load, then a lasting impression of power-supply theory, application, and operation will always be with you.

chapter nine

LOW-VOLTAGE POWER SUPPLIES

Low-voltage power-supply circuits are often used to provide operating power for transistorized circuitry or circuits using other solid-state devices. Many of these circuits require excellent regulation of output voltage, so extensive use of solid-state regulation circuitry is present in most of the power-supply circuits included in this chapter. Another type of solid-state device, the integrated circuit, contains many transistors, diodes, resistors, and capacitors all on one tiny crystal chip. Integrated circuits often require voltages of both positive and negative polarities in relationship to the circuit ground. Different solid-state devices will require different operating voltages and currents, and a broad range of low-voltage power supplies is discussed in detail in this chapter to cover many practical uses.

Generally, low-voltage supplies take the place of batteries, which were the main source of operating current for many solid-state circuits only a few years ago. Batteries inherently give excellent voltage regulation, a prime factor in the operation of most solid-state circuits. Modern, low-voltage power supplies do an adequate job of providing this stable regulation and offer the convenience of ac-driven power that never runs down.

The circuits shown in this chapter are meant to serve as guides to practical low-voltage circuit construction and use. Most of the circuit components are easily and inexpensively purchased at electronics outlets and mail-order houses. Components may be substituted to alter the output voltage, current, or regulation to values more suited to other types of electronic loads. Or one of the power supplies discussed in this chapter can serve as the basis for a more complex supply that may offer more variation of output voltages and currents. Most of the circuits can be considered noncritical, and cross-referenced solid-state devices may be used in place of the components specified with each schematic.

Although the dc output voltages generally are in the low and less hazardous range regarding human contact, carelessness in troubleshooting these circuits can lead to injury from the primary power line and, at the least, damage to rectifiers and solid-state regulator components.

Elaborate metering circuits may be added to many of these circuits, depending on what the load-metering requirements are. Chapter 6 will prove helpful for adding any further metering to the power supplies discussed here. Many of the ideas expressed and practiced in some of the schematics may be readily applied to a portion or all of the circuits in another. Experimentation is encouraged with these types of power supplies as they are, generally, a bit more complex than those of the high- and medium-voltage outputs. This is due to the absence of regulation circuits in many power supplies above the low-voltage range. All the principles and techniques applied to low-voltage supplies also apply to medium- and high-voltage supplies.

A 13-VOLT REGULATED POWER SUPPLY

Many low-voltage power supplies duplicate the voltage found in automobile electrical systems. This is usually about 13–13.5 V when the engine is running. This voltage is also a standard for powering many solid-state devices.

Figure 9-1 shows the schematic of a simple power supply that will deliver 13 V at a maximum of 2 A to any load. All the parts are standard, and no special, high-priced components will be required to complete this circuit. Transformer T_1, when connected to a 115-V power source, delivers 25.2 V of alternating current at the secondary. This voltage is rectified using a full-wave bridge circuit. The primary filter circuit is composed solely of C_1, which forms a capacitive-input filter.

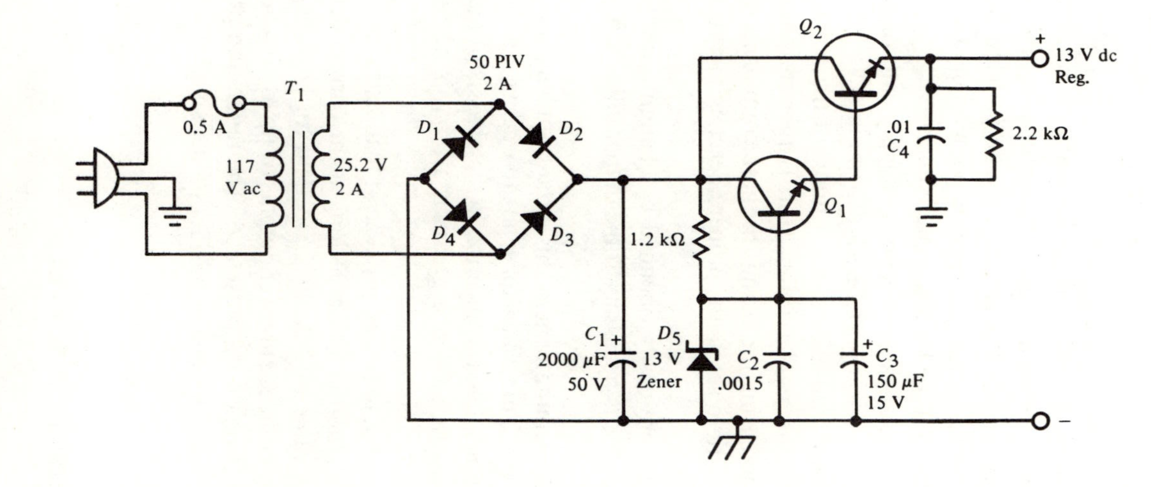

C_1 2000 μF, 50 V
C_2 150 μF, 25 V
$D_{1,2,3,4}$ 50 PIV 2 A
D_5 13 V Zener

Q_1 ECG 123 A
Q_2 2N424
T_1 25.2 V ac, 2 A

FIGURE 9-1. A 13-volt regulated power supply.

Although the regulation of voltage is good with a filter capacitor of this high value, stiffer regulation will be required for many circuits, so a simple two-transistor series regulator circuit is connected to the output of C_1. A 13-V zener diode provides a reference voltage for Q_1 to act upon in its function of providing conduction signals to the base of the series element, Q_2. The remaining components simply provide the correct operating parameters for the voltage-regulator transistors. The power supply chassis may be used as the circuit ground connection, as is shown in this schematic, or the ground may be isolated from the chassis and run to a separate jack or connection point on the front panel of the supply, depending on the builder's preference.

Maximum continuous ratings for this power supply call for no more than a 2-A current drain. During brief periods, as much as 2.5 A may be drawn. Higher currents will exceed the ratings of the power transformer, the diodes, and the regulator transistor. Larger current outputs may be obtained by increasing the ratings of these components and altering the regulator circuit to set up the proper operating parameters for its solid-state elements.

Variations in voltage are easily obtained by changing the value of the zener diode. Little else is necessary if the output voltage is to be in the range of 13 V.

The only mechanical requirement of this circuit is a heat sink for the regulating transistor, Q_2. This transistor tends to operate on the warm side in this circuit. An insulated connection to the chassis of the power supply with heat-conducting silicone grease is all that is necessary to assure the long life of this component.

This circuit may be used to power any type of load that was formerly powered by a 12-V automobile battery as long as the current demand is not over a continuous value of 2 A. It is excellent for powering small audio and transmitting devices at power outputs of up to 10W, depending on the type of electronic load presented. The regulation presented by this power supply is exceptional, considering the simplicity and the low cost of the circuitry.

TRIVOLTAGE REGULATED SUPPLY

Figure 9-2 shows the schematic of a regulated power supply with one power transformer, one rectifier circuit, and three separate regulated voltage outputs. The regulation is provided by three zener diodes

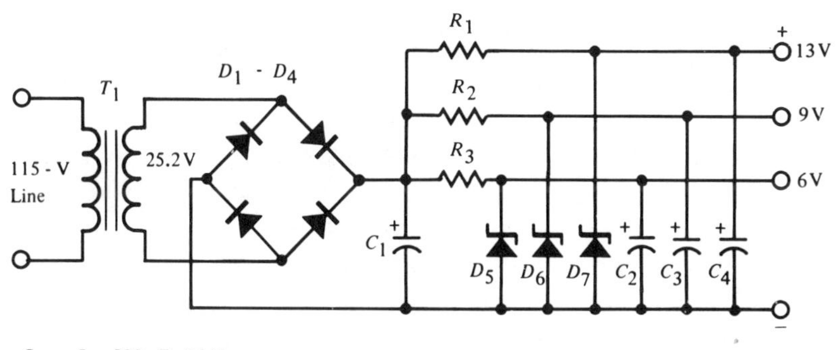

C_1 - C_4, 500 μF, 50 V
D_1 - D_4, 50 PIV, 1 A
D_5 6-V, 1-A
D_6, 9-V, 1-W zener
D_7, 13-V, 1-W zener
R_1 - R_3, 50 Ω 10 W (adjust as necessary)
T_1, 115 to 25.2 V, 1 A

FIGURE 9-2. Trivoltage supply.

connected between circuit ground and three legs of the dc output from the rectifier through three dropping resistors. This power supply provides good regulation of all three voltages when stable amounts of current are being drawn. This regulator circuit is even simpler than the one previously discussed and does not offer the same voltage stability as would be obtained with a series regulator circuit; however, for many applications, this type of power supply will offer more than adequate voltage stability. Each of the three voltage sources will deliver its design voltage of about 250 mA to an electronic load.

Secondary voltage from the power transformer is rectified by a full-wave bridge circuit and filtered by the 500- μF capacitor, C_1. This filtered direct current is then fed to a splitter circuit made up of R_1, R_2, and R_3, which forms the three separate voltage lines. The values of these three resistors are chosen to provide the correct voltage values at the output of the supply under all expected loading conditions. Values may have to be altered slightly from those specified in the schematic of Figure 9-2, depending on the type of load presented to the power supply.

Zener diodes D_5–D_7 are connected across the three legs of the dc power supply to drop excess voltage through the three resistors and regulate the final output to the load. Capacitors C_2–C_4 further the filtering process by eliminating ac ripple. Although three values of voltage are provided by this supply, more outputs could be added by taking more legs from the main voltage source through the addition of more resistors and zener diodes. This power supply was designed to deliver three specific voltages, but others may be substituted by changing the value of the zener diodes and, possibly, the dropping resistors. All other components may remain the same. Make certain that the total power drain does not exceed the transformer and rectifier ratings. Remember that some power is being dissipated in the dropping resistors as heat, and this, too, is a drain on the power supply transformer. If modifications are made, total power output to the load will be determined by the exactness of the dropping resistor values and the voltage outputs desired. The lower voltages require a larger voltage drop from the original 25–30 V dc output from the rectifiers and C_1, and, thus, more dissipation and power loss occurs in the dropping resistors.

This power supply is excellent for laboratory and bench use where many types of electronic equipment may be tested. As shown in the schematic, it can deliver three popular output voltages at moderate current levels. Many different types of electronic equipment could be adequately powered by this supply. For specific applications, it can be modified to deliver three operating voltages required by a particular electronic load. Even after the power supply is constructed, changes in voltage requirements may be adequately handled by adding another resistor, zener diode, and capacitor to the circuit to give a total of four voltage outputs (at reduced current levels), or by substituting another value of zener diode to the circuit already there.

A power supply of this types lends itself well to applications that do not require extremely stiff regulation. This is about the simplest circuit known to deliver a regulated voltage output; it has the advantage of being easy to build and, more importantly, easy to service. With so few components, circuit problems can be determined almost immediately and effective repairs made in a short period of time.

Although not shown in Figure 9-2, this power supply can be used with its mirror image, so to speak, to form an excellent integrated-circuit power supply with three voltages of two polarities, positive and negative, each. To accomplish this, another power supply is built using the same components for the first, but reversing every polarized component in the circuit. All capacitors, zener diodes, and rectifier diodes in the circuit are connected backward from the previous circuit. The output from this supply will deliver the three voltages at a negative polarity when compared to circuit ground. For this type of supply, the two power transformers can be connected in parallel to the ac power line and all components mounted on the same chassis. Many integrated circuits require voltages of both polarities to operate properly. With a power supply of this type, three different supply voltages at two polarities are available. Comparing price, simplicity, and circuit advantages, the power supply in Figure 9-2 is an excellent addition to the work bench or laboratory.

A 1.3-VOLT REGULATED SUPPLY

The schematic in Figure 9-3 shows a very compact power supply that is meant to serve as a battery replacement power source for equipment that requires 1.3–1.5 V at about 500 mA. Regulation is difficult to obtain at these low-voltage values, but the circuit shown does a very good job of maintaining the voltage level to 1.3 V over a wide range of current demands.

A small 2-A filament transformer is used to provide 6.3 V ac at the secondary, which is then rectified by a bridge circuit. The dc output

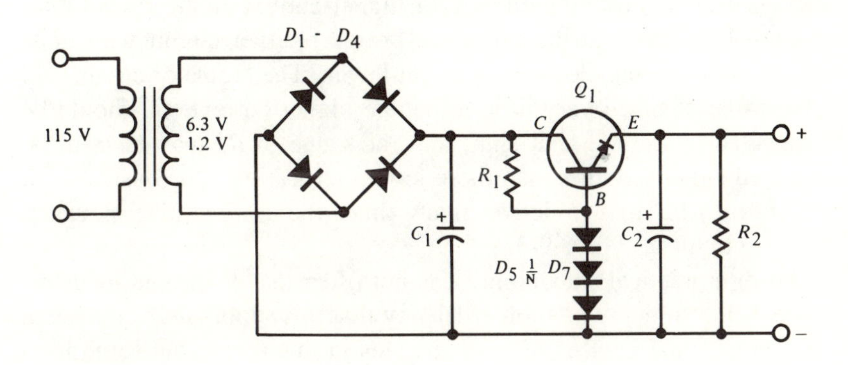

C_1, 5,000 μF, 25 V
C_2, 1,000 μF, 15 V
D_1 - D_4, 50 PIV, 1 A
D_5 - D_7, ECG - 5834 (Sylvania)
Q_1, ECG 130 (Sylvania)
R_1, 330 Ω
R_2, 1,000 Ω
T_1, 115 to 6.3 V, 1.2 A

FIGURE 9-3. A 1.3-volt supply.

from this circuit is fed to a capacitive input filter made up of C_1, a 5,000-μF, 50-V capacitor.

The voltage-regulator circuit uses Q_1 as the control element, which receives its reference from the voltage drops across the three series-connected diodes, D_5–D_7. The 330-Ω resistor connector to the base of Q_1 sets up the proper operating voltage for the transistor. A smaller capacitor is placed across the emitter Q_1 to ground and in parallel with resistor R_2, which draws current to set the minimum drain on the circuit. For these low-current applications, Q_1 does not have to be connected to a heat sink. It is designed to handle much larger current drains than this circuit is designed to supply.

The dc output voltage should be very close to 1.3 V. The three series diodes that set the reference voltage on Q_1 may present a slightly higher or lower voltage drop and may require substitution of different units until the desired voltage is reached.

The largest component in this circuit is the 5,000-μF capacitor. This one large component may be done away with, and several smaller capacitors used in a parallel configuration to give the same voltage and capacitance values. Capacitance is not critical at this point, but should be kept above 4,000 μF for best operating results.

This circuit is easily modified for higher current applications by replacing the diodes in the full-wave bridge rectifier circuit with units that are rated to handle higher current levels. The regulator circuit will have to be changed very little to handle higher current. Q_1 should be connected to a good heat sink, and the value of R_1 may have to be changed slightly. Other than these few modifications, the basic circuit shown may be used to deliver many times the design value indicated for the circuit in Figure 9-3.

From a practical standpoint, it is not often that electronic loads require 1.3 V at the regulation efficiency that this supply delivers. Often very noncritical devices require voltages in this range, but some integrated circuits will need the regulation at the voltage level that this supply is able to deliver.

DUAL-OUTPUT POWER SUPPLY DELIVERING 7 and 3.6 VOLTS DC

Integrated-circuit applications often require unusual voltage combinations depending on their application in electronic circuits. The power supply in Figure 9-4 will deliver 7V directly from the filter and uses a 3.6-V zener diode in combination with a voltage dropping resistor connected to the 7-V output to deliver a regulated 3.6 V at about 200 mA.

A small 1.5-A, 6.3-V filament transformer supplies the voltage to a full-wave bridge rectifier circuit using a 3000-μF capacitive-input filter. The approximate 8½ V delivered from this supply drops to about 7 V under a load that can approach 500 mA.

For the 3.6-V output, two 15-Ω, 1-W resistors are connected in parallel and placed across the 7-V output line for connection to a 3.6-V zener diode. A value of 7.5 Ω is needed, and combining two commonly found resistors is the easiest way to arrive at this value. Total resistor dissipation in this parallel configuration is 2 W. This dropping resistor maintains the voltage applied to the zener diode at a level slightly above 5 V. The zener then maintains the regulated 3.6-V output. Current drain from this latter output should not exceed 200 mA, which should be adequate for most integrated-circuit loads.

This circuit can be effectively incorporated into the main chassis of the electronic load because it is very small in physical size. The filter capacitor, again, is the largest component in this circuit.

Power supplies of this type are best custom designed to meet the power requirements of specific electronic circuits. If a higher voltage

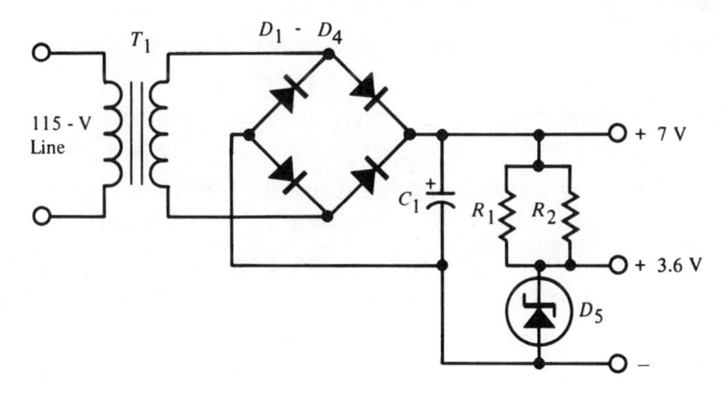

C_1, 3,000 μF, 25 V

D_1 - D_4, 50 PIV, 1 A

D_5, 3.6 V, 1 W

R_1 R_2, 15 Ω, 1 W

T_1, 6.3 V, 1.2 A

FIGURE 9-4. Dual-output regulated supply.

output is desired from the higher voltage source, another power transformer must be considered or there is the possibility of using a voltage-multiplier circuit. The same zener diode may be used if the lower voltage is to remain the same, but a different dropping resistor will be required. A value must be chosen that will drop the voltage to the zener diode to a similar level found in the circuit indicated. This particular resistor value, 7.5 Ω, was determined from the anticipated amount of current that was to be drawn by the 3.6-V load. If less than the designed 200 mA is drawn, the voltage at the zener will be higher and the device must work harder and hotter. If half the current or 100 mA is to be drawn from this supply, twice the resistance should be used. By using Ohm's law and knowing the desired voltage drop required, resistance is easily determined. In this circuit, the desired voltage delivered at the zener diode was to be approximately 4.5 V. This voltage is high enough to give the zener a value to work with while not being excessively high. At 5.5 V to the zener, the voltage will not drop below 3.7 V even when line voltage is low. If the value drops below the zener diode's rated voltage, regulation ceases.

Using this line of thought and assuming that a current drain of 50 mA is desired from the 3.6-V supply, the value of the resistor is found by dividing 1.5 V by 0.050 A (50 mA). The answer is 30 Ω. The 1.5 V is the desired voltage drop from the main 7-V supply to deliver approximately 5.5 V at the zener diode.

If the main portion of the supply is set up to deliver 12-V output under load and a 3.6-V lower output is still desired, a voltage-drop figure of 6.5 V is used and divided by the amount of current to be drawn from the lower supply. A voltage of 5.5 V will still be delivered to the zener diode for a circuit such as the one described if a dropping resistor of 130 Ω is used, assuming a drain of 50 mA from the 3.6-V supply.

A zener diode with a different voltage drop may be used to deliver other values of voltage from the regulated circuit. A working voltage of a value that is approximately 15–20 percent higher than the regulated output is desired and may be considered when computing the drop from the source voltage line.

A 4- TO 20-VOLT VARIABLE POWER SUPPLY

Continuously variable voltage is often necessary in many power supplies to operate various pieces of electronic equipment. Transmitting loads such as small oscillators may be adjusted as to power output by simply raising or lowering the main voltage supply. With a continuously adjustable supply, these power levels can be exactly set to the proper operating levels.

Figure 9-5 depicts an adjustable supply that can be set for a minimum of approximately 4 V to a maximum of about 20 V and every value in between. A small integrated circuit is used to accomplish the regulation function and is coupled with a small variable resistor to enable voltage adjustments to be made. This very simple power supply exhibits excellent regulation of dc voltage and easy construction and wiring techniques.

Owing to the current limitations of the integrated circuit used in this supply, low current levels of no more than 180 mA must be drawn by the electronic loads. Integrated circuits (IC's) require special consideration when making solder connections as they are very sensitive to heating. Improper soldering techniques can lead to total damage of these devices. A low-wattage, pencil-type soldering iron should be used with a power rating of no more than about 30 W. Each lead of the

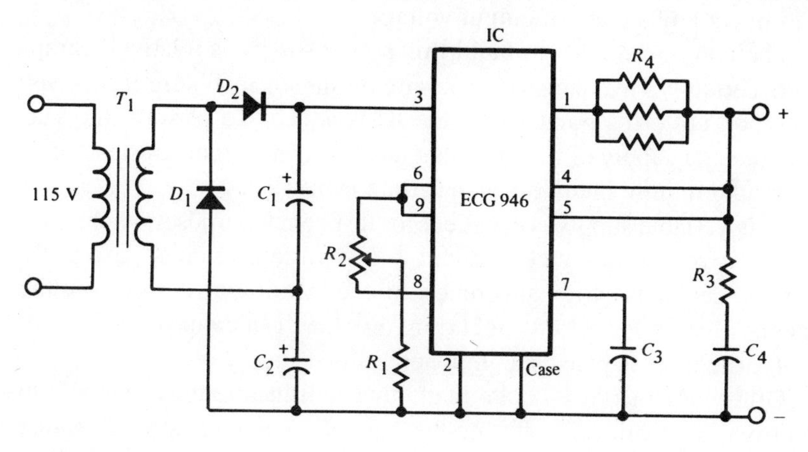

C_1 C_2, 3,000 μF, 15 V
C_3 C_4, 1 μF, 50 V
D_1 D_2, 50 PIV, 1 A
IC, ECG 946 (Sylvania)
R_1, 6.8 kΩ
R_2, 50 kΩ liner taper control
R_3, 25 Ω ½ W
R_4, 3.3 Ω (three - 10 Ω, ½ W resistors in parallel)

T_1, 12.6 V, 500 mA

FIGURE 9-5. Variable power supply.

IC should be grasped firmly with a pair of pliers just above the joint to be soldered to form a protective heat sink when the iron is applied. This prevents large amounts of heat from traveling through the wire leads to the heart of the encased circuit and causing damage to the solid-state interior.

The circuit begins as a full-wave voltage doubler made up of D_1 and D_2 in connection with capacitors C_1 and C_2. The no-load dc output from this portion of the power supply is approximately 21 V. This output is fed to the IC with the positive connection at pin 3. The IC delivers the output voltage through series resistor R_4, which also works with the IC to form a current-protection circuit. Whenever the leads of the regulated output are shorted together, the power supply automatically shuts down all current flow. This prevents damage to the solid-state components in the circuit. Capacitor C_4 and resistor R_3 form a high-frequency compensating network for added stability.

Voltage adjustment is accomplished by the variable control R_2, which is wired in parallel with pins 6 and 8 of the IC. Make certain that this 50-k Ω potentiometer is of the linear taper variety to assure smooth and linear adjustment of output voltage.

The components list to build this power supply is relatively simple and, certainly, inexpensive. The circuit shown in Figure 9-5 is very common up to the point where the IC is added. To perform the same functions, a supply of this type that used discrete circuits instead of the IC would require many times the components listed here.

This variable supply is excellent for test bench purposes where many different voltages may be required to test various circuits. Use of this supply demands strict adherence to the maximum current requirements. Even brief periods of current overload can damage the IC, possibly, the most expensive component in this supply.

Other varieties of IC's are made that will handle much higher current outputs while still offering the regulation and adjustment features found in this supply. Integrated circuits are available in many different current and voltage ratings, as are other solid-state components that are offered as discrete devices. Special operating requirements should be considered, and then integrated circuits can be studied and chosen on their adaptability to the circuit values desired.

REGULATED 9-VOLT POWER SUPPLY

Nine-volt power supplies are not often required in the operation of many laboratory types of electronic loads. Some equipment may require this voltage or, at least, will operate satisfactorily at this value, but the main use for 9-V regulated supplies is to power small entertainment devices such as transistor radios, tape recorders, digital clocks and calculators, and many more. Figure 9-6 shows a very simple 9-V power supply using a series pass regulator circuit to provide good voltage stability, while requiring only four additional components over what would be needed to construct an unregulated power supply at the same voltage.

Referring to Figure 9-6, the power transformer is rated at 25.2 V across the entire secondary. A full-wave center-tapped rectifier configuration coupled with the capacitive input filter delivers a dc output under no-load conditions of approximately 17½ V. This value is delivered to the series-pass regulator circuit.

The control element in this circuit, Q_1, receives its reference voltage value from a 9.1-V zener diode. Resistor R_1 sets up proper operating parameters for the regulator circuit. Output voltage should be very stable under all loads up to a maximum of approximately 260 mA. Q_1 should be mounted on an insulated heat sink to assure continued operation within dissipation limits.

As was previously mentioned, the power supplies described in this chapter can be greatly modified to better mate with the type of electronic equipment that will serve as the power-supply load. By making

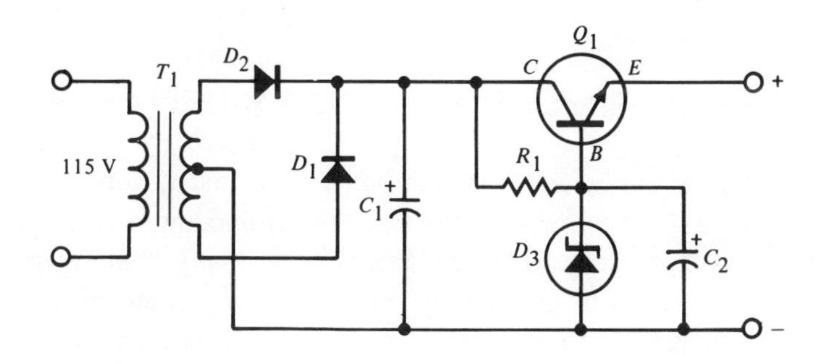

C_1, 500 μF, 25 V
C_2, 100 μF, 25 V
D_1 D_2 50 PIV, 1 A
D_3, 9 - V zener diode
Q_1, ECG 124 (Sylvania)
R_1, 560 Ω, ½ W
T_1, 12.6 V 1 A

FIGURE 9-6. A 9-volt power supply.

minor changes to the regulator circuit components, this power supply may serve as a well-regulated 6-, 12-, or 15-V supply. Other modifications may be necessary to arrive at the 9-V output when specific parts are not available. Zener diodes may have to be connected in series if a 9.1-V unit is not available. Capacitors may be paralleled in the circuit to match the capacitance of the ones specified. In some areas, 25.2-V power transformers are not as easily obtained as the 6.3- or 12.6-V models. This is really no problem, because any of these transformers may be used with the same regulator circuit to deliver the rated 9-V dc output. If a 6.3-V power transformer is available, a voltage-doubler circuit of the full-wave design should be used. If a 12.6-V transformer is used, a full-wave bridge circuit will deliver the correct values to the regulator circuit.

In the description of the operation of this power supply, it was stated that the full-wave center-tapped rectifier and filter combination delivers approximately 17.5 V to the regulator circuit. Theoretically, this value would be 17.64 V according to the formulas studied earlier. By computing the anticipated no-load output from each of the alternate power transformer circuits just discussed, it will be noted that the output of these supplies to the regulator is exactly the same as the circuit shown in the schematic of Figure 9-6, or 17.64 V dc. This value does not have to be exact, and slightly higher or lower dc voltages to the regulator circuit may be used with the same results.

A 0- 25-VOLT VARIABLE SUPPLY AND BATTERY CHARGER

Many bench applications still require batteries in some form or other, especially in portable field-operated equipment. The rechargeable nicad batteries are popular and practical for many of these applications, and can be recharged many times before replacement is necessary. The battery charger pictured in Figure 9-7 is basically a low-voltage power supply with a transistor voltage output control circuit. Variable control R_1 sets the base current on the control transistor determining the amount of current conduction through the transistor.

The power transformer is a 25.2-V, 1-A unit that supplies voltage and current to the half-wave rectifier circuit and the capacitive-input circuit made up solely of capacitor C_1. R_1 forms a voltage-divider network for control of the base of the transistor. A 0- to 35-V dc meter monitors the output voltage to determine proper settings for the nicad batteries under charge.

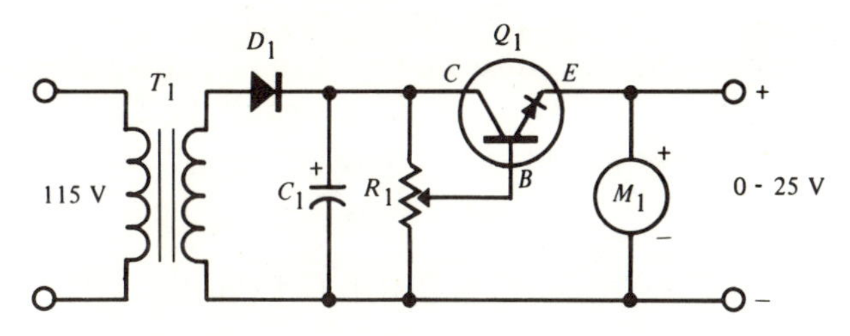

C_1, 500 μF, 50 V
D_1, 200 PIV, 1 A
M_1, 0 - 35 V DC Voltmeter
Q_1, ECG 184 (Syvania)
R_1, 2,500 - Ω linear taper control
T_1, 25.2 V 1 A

FIGURE 9-7. Variable power supply battery charger.

This power supply is not regulated by any electronic means, and the voltage variation will render this supply unsuitable to directly power any electronic load that requires stiff regulation. Other devices, though, that are not as critical as to the stability of input voltage may be powered with this supply. Small dc motors may be used very effectively for speed-control purposes using this circuit.

For battery-charging purposes, the nicad cell is placed across the power-supply contacts with the voltage control in an off or low position. The voltage is then adjusted to a point slightly above the rated voltage value of the battery to be charged. Charging time will be determined by the size and charging rate of the battery.

It is necessary for transistor Q_1 to be mounted on a good heat sink, because, during periods of use at very low voltage outputs, the case will become very warm and this heat must be dissipated. The collector of this transistor is electrically connected to the metal case, so an insulating washer, which is usually supplied with this component, must be used to prevent a short circuit with the chassis or circuit ground. Alternatively, the transistor may be mounted directly to an external heat sink, which, in turn, is completely insulated from the chassis. Make certain that an adequate amount of silicone grease or some other heat-conducting compound is used on the bottom surface of the transistor case, on the insulating washer, if used, and on the heat sink or chassis where the transistor is to be mounted. This grease helps conduct heat from the transistor to the case of the heat sink. It does not conduct electricity.

Another use for this variable supply would be to obtain regulated voltage by adding external, discrete regulating circuits for various voltages. The input voltage to these circuits could be accurately set by using the meter at the output of the variable supply circuit.

A 28-VOLT SUPPLY USING A SHUNT REGULATOR

Many of the electronic regulator circuits used with power supplies discussed so far in this chapter have used series-pass regulators to maintain the voltage value at an almost constant level. This section will deal with a dc power supply with an output of 28 V, plus or minus 0.14 V. This regular circuit uses a two-transistor, shunt-regulator circuit that does a very adequate job of providing good regulation up to current outputs of 400 mA.

Referring to Figure 9-8, it can be seen that a bridge rectifier is used to supply dc output to the capacitance-input filter. The power transformer can be supplied with either a 36- or 40-V secondary winding. The main concern is to supply at least 44 V dc to the regulator under loading conditions. Variations in rectifiers may be used to accommo-

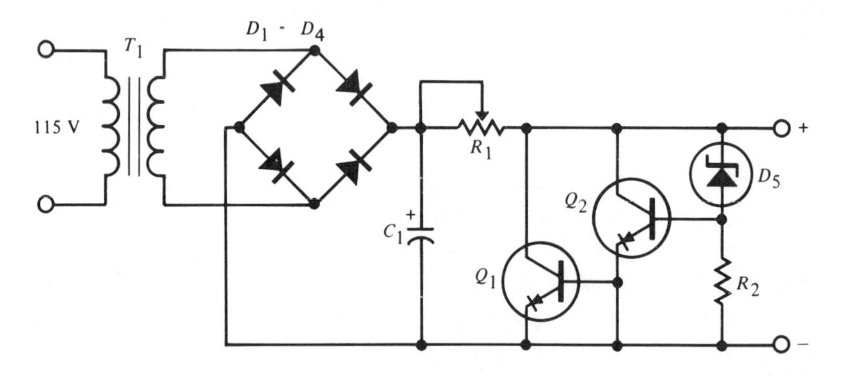

C_1, 1,000 μF, 200 V
D_1 - D_4, 400 PIV, 1 A
D_5, 27-V ½-W zener
Q_1, ECG 175 (Sylvania)
Q_2, ECG 128 (Sylvania)
R_1, 75 Ω, 50 W adjustable resistor
R_2, 1,000 Ω, ½ W
T_1, 30 - 40 V, 1 A

FIGURE 9-8. Shunt regulator supply.

date transformers with other secondary windings. For example, a 12.6-V winding may be used with a full-wave voltage-tripler rectifier circuit to produce a no-load output of approximately 50 V. Any value of voltage from 44 to 54 V will be satisfactory.

The voltage-regulator circuit is of the shunt design and uses a 27-V zener diode connected to the control transistor, which is, in turn, connected to the control element, transistor Q_1. R_1 serves as a dissipation source for the voltage that is dropped to produce the regulated output. When voltage output begins to exceed the 28-V design value, the zener diode senses this and drops a small portion of this output, which triggers Q_2, Q_2, in turn, sends a driving signal to the control element, Q_1, which conducts to a point where voltage is dropped across the series resistor R_1. Thus regulation is maintained under conditions of varying loads. The other component in the circuit, R_2, sets up proper operating parameters on the base of Q_2 and acts as a dropping resistor to initially trigger Q_2 in connection with the zener diode.

Regulation efficiency with this circuit is very high. Voltage is maintained to a value that is within 0.5 percent of the design value, plus or minus. Owing to the high, unregulated input voltage, dissipation of the series resistor R_1 and the control transistor Q_1 tends to be high. Adequate heat sinking is absolutely necessary for Q_1, and a very large resistor with a power dissipation rating of at least 30 W is required for R_1 to maintain an acceptable safety factor for this device.

This shunt circuit does not give as good regulation efficiency as would a series pass circuit; but for a simple circuit offering medium to high amounts of current output, it proves to be very acceptable. Only in powering electronic loads that require laboratory-quality stability of voltage value would this power supply not prove adequate.

The regulator circuit consumes only a small amount of space when completed and may easily be mounted in the same case or on the same chassis as a power supply in the 44- to 55-V range to provide regulation desired. The heat sink used with Q_1 must be insulated from the chassis or circuit ground, or a short circuit will exist and, probably, destroy Q_1 and other solid-state components in this circuit.

COMPUTER POWER SUPPLIES

Power supplies for computer circuits cover such a large range that an entire text could easily be written on the subject. Computer power supplies, depending on the service, are usually of the low-voltage regulated type studied in this chapter. Integrated circuits are used extensively and in many stages throughout the regulation circuit to provide regulation efficiency, which is not usually required for most other electronic purposes.

Some computer circuits also require power in the medium- to high-voltage level, again with very good regulation characteristics. Special circuits are used to regulate this voltage to a high degree at costs that do not make them practical for most applications. Sometimes unusual input voltages are used at frequencies of other than 60 Hz.

Owing to the extreme efficiency required, many computer power supplies contain circuitry at the input to the power transformer to regulate the source voltage. This, in turn, helps to keep the resulting dc output at a better level of regulation.

Home computer systems have come of age just recently. These circuits use power supplies on the order of the better-regulated circuits shown in this chapter. Usually, solid-state components of the discrete type are done away with in favor of using integrated circuits exclusively where possible. Instead of using four separate diodes in a bridge rectifier assembly, one IC that contains four diodes in the bridge rectifier hook-up pattern is used. This one small chip saves space, but, more importantly, provides a more secure and dependable mechanical connection that is less subject to temperature and shock variations. Integrated circuits that contain their own resistors, capacitors, and diodes, as well as transistors, thyristors, zener diodes, and other solid-state devices all on one small chip in one tiny plastic or metal package are also used for the main regulating portion of the power supply. Computer circuits, although complex, may draw only meager amounts of current and on a steady basis. This makes the task of supplying good voltage regulation a bit easier. Higher current levels often necessitate the use of discrete power-handling components, which require more labor in obtaining acceptable regulation.

It can be seen that regulation of the input as well as the output voltage is extremely important for computer operation. Regulation, in fact, is the key element for such circuits. Poor regulation in most electronic applications, especially for hobby purposes, may mean only the inconvenience of tracing the trouble and changing components in the power supply to correct the situation; at worst, the inconvenience of delay is the only problem that may be caused. But in computer circuits, a regulation problem may cause actual damage to hundreds of circuits and components; even worse, whole banks of information could be damaged or wiped clean at a cost of millions of dollars in replacement expense and time.

Equipment that is used for medical purposes may also require the regulation efficiency that computer circuits demand. Due to the critical nature of many types of electronic monitoring devices, failure could cause serious complications. Here power-supply dependability and regulation are of prime concern, as well as the remainder of the electronic circuit. Backup power supplies are often included in these types of equipment to automatically take over if the main internal power supply fails.

Generally, the more complex and critical the electronic load is, the greater the need for high regulation efficiency in dc power supplies. Also, the dependability factor of components becomes critical, and components may be designed for these specific circuits that are specially manufactured to closer tolerances than the normal commercial parts available to the general public. This special manufacturing usually dictates a cost for components that may be many, many times that of standard components; but the added cost is a very insignificant factor when compared to the tremendous expenses that are involved in an equipment failure.

QUESTIONS

1. In a simple zener-diode-regulated power supply, what is the purpose of the series resistor?
2. What is a VR tube?
3. Many low-voltage power supplies take the place of what electrical component?
4. What will result if adequate heat sinking is not provided for the series control element in a series-pass regulator circuit?
5. Why is good regulation required from many low-voltage power supplies used for electronic equipment?
6. In a power supply with no electronic regulation circuit, what could cause large variations in the output voltage under various loads?
7. In a circuit with electronic voltage regulation, what conditions

could cause a large fluctuation of output voltage under various loading conditions?

8. What devices besides transistors may be used to electronically regulate output voltage?
9. A series-pass regulator circuit that uses two transistors in series with the dc line does so for what purpose?
10. What is meant by the term *regulation efficiency?*

chapter ten

MEDIUM-VOLTAGE POWER SUPPLIES

Medium-voltage power supplies are in great use today to power many types of electronic equipment, several of which use tubes exclusively or in part and need these higher voltages to set up proper operating parameters for these electronic components. Medium-voltage supplies may be used to provide anode, grid, and screen grid voltages for tubes operated in the load circuit. Whereas the anode and screen grid of a tube usually requires dc voltage that is positive in relationship to circuit ground, control grids often require voltage that is negative to ground. Medium-voltage output from a power supply can usually be thought of as covering a value range from a low of about 40 V to a high of 600 or 700 V depending on circuit requirements.

Many medium-voltage supplies may be called on to deliver two, three, or even more voltages, some of which may be of different polarities. In these cases, power transformers that have several different secondary windings may be used in power-supply design and construction. Medium-voltage supplies may provide all the operating voltages required for a particular electronic load or may provide only a few. High-power circuits often require medium and high voltage to operate. Here the medium-voltage supply may be used in conjunction with a separate high-voltage supply, the latter providing a high value of anode voltage for tube circuits.

Electronic regulation of voltage output is not used as much with most medium-voltage supplies as it is with many of the low-voltage supplies studied in Chapter 9. Control and screen grid circuits will often require a relatively stiff source of voltage, and some form of electronic regulation may be necessary for these types of supplies. As is usually the case, electronic load considerations will determine this factor, as well as current ratings of various supplies.

In transmitting circuits that also contain an internal receiver, medium-voltage power supplies often provide all but the anode voltage for operating the transmit section, while supplying all the voltages to operate the receive circuits. Operating two circuits in a transceiver may not call for an extra large current rating on the medium-voltage supply, as many of these circuits switch the power supply feeds from transmit to receive as each of the sections of the load is called into service. This drops the load current requirements, because the transmit section is inoperative while the receive circuits are activated, and vice versa. An ac filament voltage may also be provided by a medium-voltage power supply that has another secondary winding for this purpose. Filament supplies can use either alternating or direct current for most applications, and, when possible, alternating current is usually opted for to avoid the added costs necessary when providing another rectifier and filtering circuit.

Medium-voltage power supplies are required in almost every circuit that uses tubes; therefore, there are more medium-voltage power supplies in use today than any other type of supply. Even many transistorized circuits use tubes in the power stages. The low-voltage supplies normally used with transistorized circuits do not provide voltage and current outputs that are applicable to tube circuits, and a separate medium-voltage supply must be used.

With several secondary windings in use to power electronic equipment, it should be remembered that the primary winding will draw a little more power, owing to losses, than the total power drawn from all secondary windings in use. Some power transformers are designed to deliver many types of voltages with many secondary windings, but some of these transformers do not have primary windings able to tolerate the amount of current that must be drawn in this section should all the secondary windings be used at once and at their maximum ratings. This type of transformer will usually include a placard stating that only three secondary voltages may be used at one time, or provide other information regarding the total amount of power that may be safely drawn from the combined windings in the secondary.

Special components are commonly available for multivoltage supplies. Some capacitors offer several different values of capacitance and voltage ratings in one small unit. Actually, several capacitors are connected to a common ground point and confined to one metal case, with insulated contacts for connection to the positive side of each unit. One section may be rated at 450 working volts and 40 μF; another may have a 250-working-volt rating and a capacitance of 15 μF. A capacitor with these ratings might be suited to a power supply delivering output voltages of 350 V dc and 150 V dc to the load. These multisection capacitors are available in many different voltage and capacitance configurations and can usually be purchased for many varying values of output voltage. Their main advantage is in compact size. A three-section capacitor that contains three different capacitance values is often not appreciably larger than a single capacitor. The cost of this type of capacitor is, likewise, not much more than a single capacitor, and much less expensive than purchasing three discrete components.

When a positive and negative voltage are required from the same power supply, one secondary winding can be made to furnish both. Figure 10-1 shows how a bridge rectifier can be used to deliver positive output voltage from a transformer secondary in the usual manner. The center tap on the secondary winding is not normally used with a bridge rectifier circuit, but here it is providing the power-supply ground point.

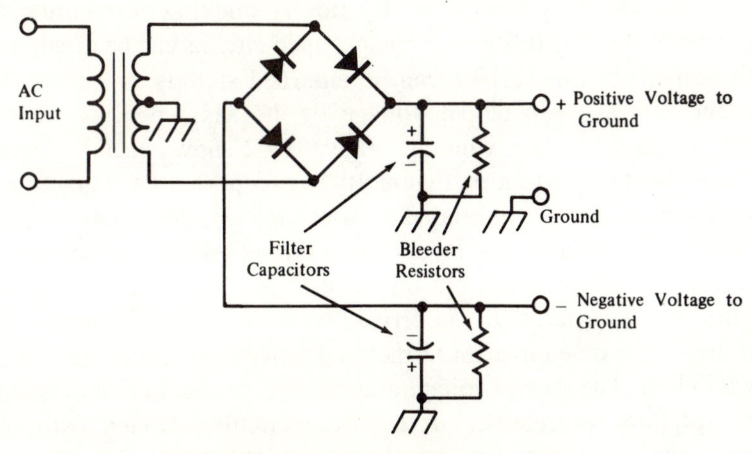

FIGURE 10-1. Bipolar power supply.

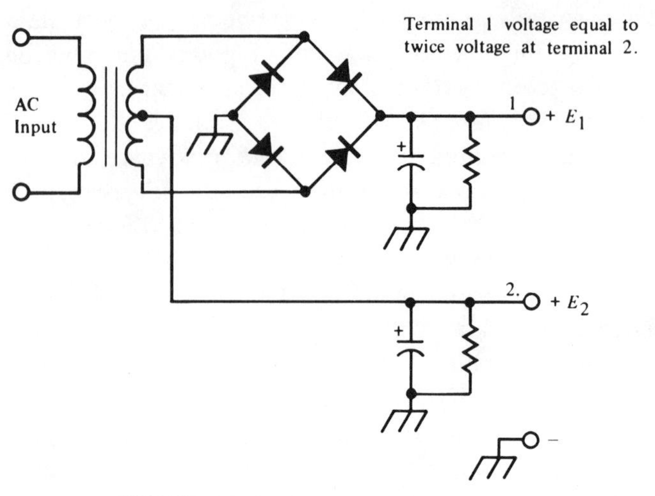

FIGURE 10-2. Multivoltage supply.

The lead that is taken from the back of the bridge rectifier circuit provides negative voltage in relationship to the center-tap ground. Although the rectifier configuration looks like a standard bridge circuit, it is actually two full-wave center-tapped rectifier circuits combined. The output voltage will be equal to the dc equivalent of half the total secondary ac output from the transformer, because the center tap is used. The voltage lead coming from the front of the rectifier stack is the positive lead with respect to ground. Notice the polarity markings on the electrolytic capacitors. Both voltages are identical in value, but one is positive and the other is negative. By closely studying the rectifier circuit, the two discrete full-wave center-tapped circuits can be discerned.

The circuit in Figure 10-1 can be modified slightly to supply two different values of dc output voltage of the same polarity from a center-tapped secondary winding. Figure 10-2 shows such a circuit, which is identical to that of Figure 10-1, except that the transformer center tap is not the circuit ground but a voltage lead, which operates as another positive source in relationship to circuit ground, which is now found at the lead coming from the back of the bridge rectifier array. The first voltage source, E_1, is derived from the full-wave bridge rectifier and is the dc equivalent to the total ac voltage across the secondary winding. The second voltage source, E_2, is equal to half of E_1 and is derived from two rectifiers in the bridge circuit and the transformer center tap. E_2 is rectified by a full-wave center-tapped circuit that is made up of two of the rectifiers in the bridge circuit. This dual-voltage power supply uses full-wave bridge and center-tapped rectification simultaneously.

This type of circuit functions best with loads that do not require isolation of the two voltages provided by the supply and whose regulation requirements are not high. Because both voltages are derived from the same transformer secondary winding and from the same rectifier assembly, when one voltage source is subject to a surge demand for higher current, voltage will tend to drop at both leads, E_1 and E_2. If E_2 has a value of 600 V, E_1 will be approximately equal to half that value, or 300 V. The total power drawn from both voltage sources should not exceed the power rating of the transformer secondary. Although capacitor-input filtering is used at the output of each voltage source, this is not mandatory. One source could use a choke-input filter while the other could use capacitor input with a separate, electronic regulator circuit. For the most part, the two voltage sources can be treated as separate power supplies, regarding filtering and regulation.

The next section will deal with specific power-supply circuits with voltage outputs within the ranges already discussed. Each circuit will be explained as to operation and performance. Component ratings are given for the specific voltages and currents produced by each supply, but circuit alterations are easily accomplished to provide other values by changing the proper components. These schematics can serve as a guide for building a myriad of different supplies that operate in a like manner.

DUAL-VOLTAGE SUPPLY

The power-supply circuit in Figure 10-3 has a high-voltage output of 750 V under load, and a low output from the second source of approximately 275–300 V under load. Current ratings are 300 mA for the higher voltage and 125 mA for the lower.

Referring to the schematic, input is derived from a standard 115-V source, which is fed to the transformer primary through a protection network consisting of relay K_1 and a 50- Ω resistor. When switch S_1 is thrown to the *on* position, a voltage drop is induced in the circuit because the line current is drawn through the 50-Ω resistor for the fraction of a second that it takes K_1 to engage its contacts, shorting out the series resistor. The coil of the relay is in parallel with the ac line and

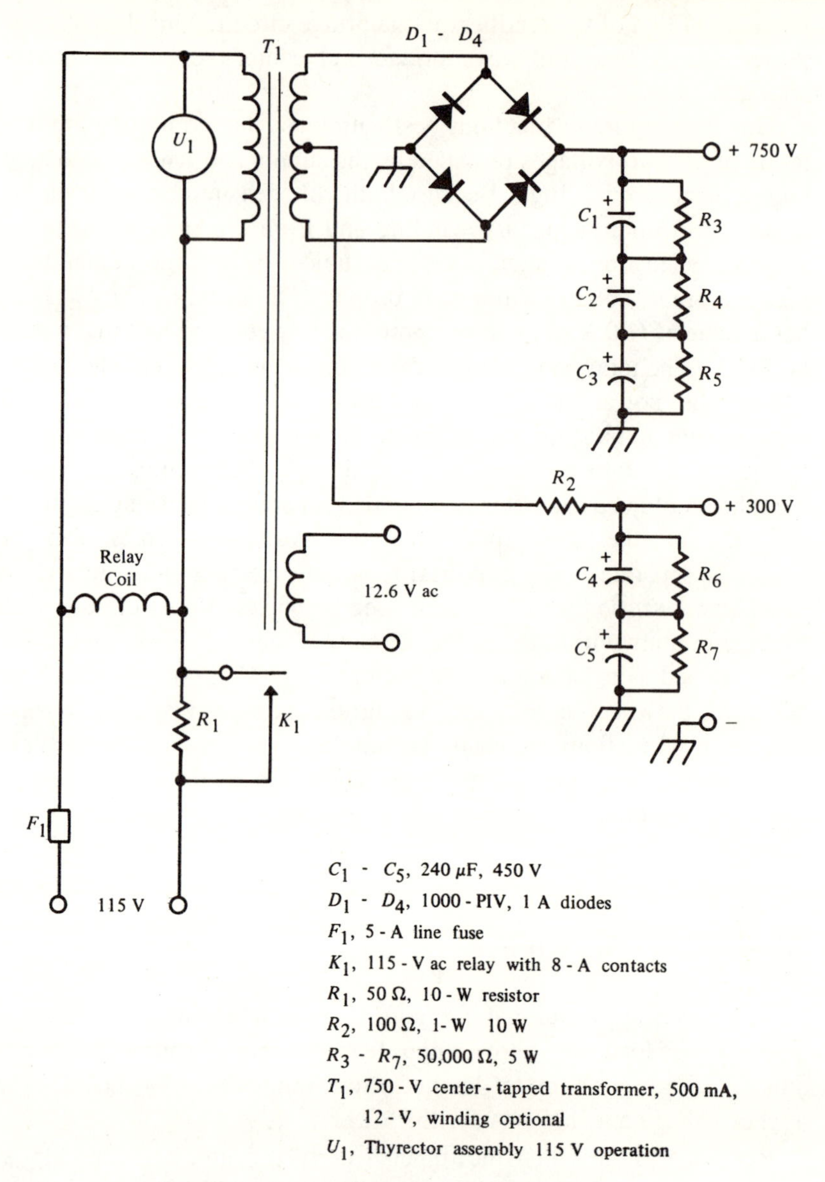

FIGURE 10-3. Dual-voltage power supply.

begins to trigger when the switch is thrown. This delay in applying full voltage to the transformer primary, although very short, allows enough time for the filter capacitor bank to partially charge. This relay action prevents the sudden surge that occurs when electronic equipment is first energized from damaging the rectifiers.

Power transformer T_1 has two secondary windings, one with a 750-V output that is center tapped; the other is rated at 12 V ac and is used to supply filament voltage to a tube circuit. An economy power-supply circuit is used for rectification, as was discussed previously in this chapter. The bridge rectifier circuit supplies the higher voltage to the capacitor bank; the center-tapped full-wave circuit provides the lower voltage to capacitors C_4 and C_5, which make up its filter network. Resistors of 50,000 Ω are placed in parallel with each of the filter capacitors to bleed off stored current when the supply is shut off. A thyrector assembly is used across the primary windings of T_1 to suppress line voltage transients that may occur and damage the power-supply components.

This is a very simple and very efficient dual-voltage supply that offers a reasonable amount of power output in a compact package. The components are usually mounted on an aluminum chassis and then covered with a safety shield to prevent accidental contact. The wiring is straightforward and noncritical. Although commercial transformers can be ordered for this type of circuit, many of the older black and white televisions have transformers that very nearly match the one described in this circuit, and that may be available for considerably less cost outlay than a new unit ordered from the factory.

The no-load output voltages from this power supply will be approximately 1,000 V for the higher value and about 500 V for the lower. Under loading conditions these values will drop to levels closer to the nominal values presented with the schematic. The capacitor bank for the 750-V leg of the supply is rated at 1,350 V, and C_4 and C_5 are rated at a total of 900 V in order to prevent capacitor breakdown during periods of higher voltages due to the lack of current demand from the load. The negative lead for this supply is the chassis itself, although the chassis connection made at the back of the bridge rectifier circuit may be isolated from ground if desired. In this case, all ground connections indicated by the schematic should be removed from the chassis and connected directly to the lead coming from the back of the rectifier stack, which is now the circuit ground.

This power supply will draw approximately 3A of alternating current when operated at maximum ratings. A 5-A fuse is used to protect the circuit and allows a margin for sudden surges that may be drawn from the supply, while providing an adequate safety factor to prevent damage to the larger components.

If the 12-V filament winding on the power transformer is not needed to supply operating voltage and current to tube filaments, this voltage may be rectified and used to control dc relays, or even to power an indicator light that signals activation of the power supply. A voltage doubler, tripler, or quadrupler rectifier stack could also be used to multiply the output voltage of this winding to a level that could supply control grid voltage to a tube-type circuit.

One final note: the transformers that are often used with this circuit tend to operate at a relatively high temperature when maximum load current is being drawn. This is a normal operating condition and should be of no concern unless the temperature becomes extreme.

TWO-TRANSFORMER DUAL-VOLTAGE SUPPLY

Figure 10-4 shows a different type of dual-voltage power supply that uses two transformers to supply the desired voltage outputs. An unusual feature of this supply is the manner in which the second transformer is used. This supply will deliver about 325 V from the main transformer, under load, and 0–130 V negative from the second trans-

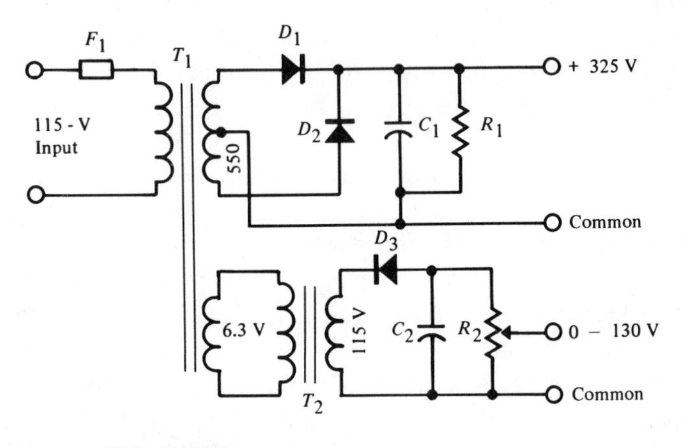

C_1, 200 μF, 450 V
C_2, 250 μF, 250 V
D_1 D_2 D_3, 1000 PIV 1 A
F_1, 3 - A line fuse
R_1, 50,000 Ω, 5 W
R_2, 50,000 Ω 5 W Linear Taper
T_1, 550 - V center tap, 350 - 400 mA, +63 V, 1 A
T_2, 115 V to 6.3 V, 1 A (reversed)

FIGURE 10-4. Two-transformer supply.

former and associated circuitry. Current ratings can be close to 400 mA for the higher voltage, depending on the transformer rating, and about 50 mA for the lower voltage. The low-voltage supply is designed to deliver grid current to tube circuits. This supply does not need to have a higher power capability, because grid circuits usually draw very little current. A method of adjusting this lower supply voltage is provided by the variable resistor.

Alternating current is applied to the transformer primary in the conventional manner. The higher voltage winding of T_1 feeds a conventional full-wave center-tapped rectifier circuit. The center tap of the transformer is connected directly to the chassis, which serves as a common ground point for this supply. The unfiltered direct current is smoothed out by a single capacitor, C_1, which has R_1, the bleeder resistor, wired in parallel with its contacts. The resultant output will be approximately 385 V with the transformer designated in the parts list. This value will drop to about 325 V under a light load, to a little less than 300 V for heavy drains on supply current.

T_2 is connected in an unusual manner to provide output bias voltage that is negative in polarity with relationship to the chassis ground. T_2 is a 6.3-V, 1-A filament transformer with a 115-V primary; but in this application, the primary winding serves as the secondary and the secondary as the primary. The 6.3-V winding of T_2 is connected to the 6.3-V secondary winding of T_1. When the circuit is activated, the 6.3 V at T_1 provides primary power for T_2, which has an output of 115 V when connected in this manner.

Transformers are reversible devices. A transformer that has a primary voltage of 230V and a secondary of 2,000 V may be reversed if so desired, and a source of 2,000 V ac applied to the secondary winding will produce a value of 230 V at the former primary windings. Therefore, it can be seen that the secondary and primary windings of a transformer are really determined by how they are used in a circuit. Regarding T_2 in this circuit, normally the 6.3-V winding would be called the secondary, but the reverse is true in this case. Current ratings still remain the same in each winding whether used as secondary or primary. T_2 is rated to deliver 6.3 V at 1 A. With the primary now acting as the secondary, the 6.3-V primary may now draw 1 A from its ac source and deliver approximately 6 W of power at 115 V from the secondary at 55 mA of current.

The rectifier circuit of T_2 is a simple half-wave arrangement with a capacitive-input filter composed of C_2. R_2 is a 5-W, 50,000-Ω linear taper control that can set the bias voltage at any point from zero to a maximum output of 130 V. Bias voltage must often be set on tube circuits during actual operating conditions with all voltages applied. The bias voltage control is then set to deliver the desired voltage to the tube grids.

Although not shown in this schematic, a power transformer that has another winding might be desirable, depending on the application intended for the supply. Another 6.3- or 12-V winding would provide a filament voltage source for the electronic load. A separate filament transformer could also be used with its 115-V primary winding connected to the ac main line on a separate switch that would allow the filaments to be activated before the higher voltages are applied. Some electronic circuits require that the tube filaments be allowed to heat for a minute or more before the other operating voltages can be safely applied.

The power supply just described is usually suited to low-power circuits such as tube-type receivers, audio amplifiers, low-powered transmitters, and like equipment that uses tubes with operating voltages in the range provided. Components for this supply are very common and can be found in almost any electronics supply store or warehouse. A used television transformer may be called to service again in this circuit if the proper voltage can be found. Although a 550-V secondary is called for, a secondary voltage of 600 or 650 V will be suitable for use in the circuit without changing any of the other components. The dc output voltage will be higher but still within the maximum voltage ratings of the components used.

If a transformer with a higher secondary voltage is used and the higher dc output is a problem, a resistor with a 20-W dissipation rating and a value of 100 Ω may be placed in series with the center tap and chassis ground. When current is drawn from the supply, the resistance in the circuit will drop the dc output voltage. A 400-mA drain will create a drop of about 40 V. The voltage drop is equal to the ohmic value of the resistor times the output current, $E = IR$.

A 700-VOLT SUPPLY WITH HALF-POWER SWITCH

Figure 10-5 shows a power supply that has a nominal output of 700 V under load currents of approximately 300 mA. This circuit is straightforward and the same as an earlier circuit discussed from the secondary of the power transformer to the output terminals. The main difference lies in the primary winding, which is designed to operate at 230 V ac or 115 V ac. The primary consists of two 115-V windings that are connected in series for 230-V operation. Although 115 V will operate this transformer at full secondary output if the primary windings are connected in a parallel configuration, this circuit depends on 115-V operation to provide a means of halving the dc output voltage and a 230-V input for a normal output of 700 V.

A bridge rectifier and capacitive-input filter circuit provide approximately 700 V of dc output at up to 300 mA while the input voltage is at the higher level. When the half-power switch S_2 is thrown, the dc output voltage drops to approximately 350 V with a maximum current rating of the same 300 mA to the load.

This half-power modification is accomplished solely at the input to the transformer primary windings. In the normal, high-output position, the switch applies 230 V to the windings. When thrown to the low-power position, the switch connects the neutral lead to one side of the

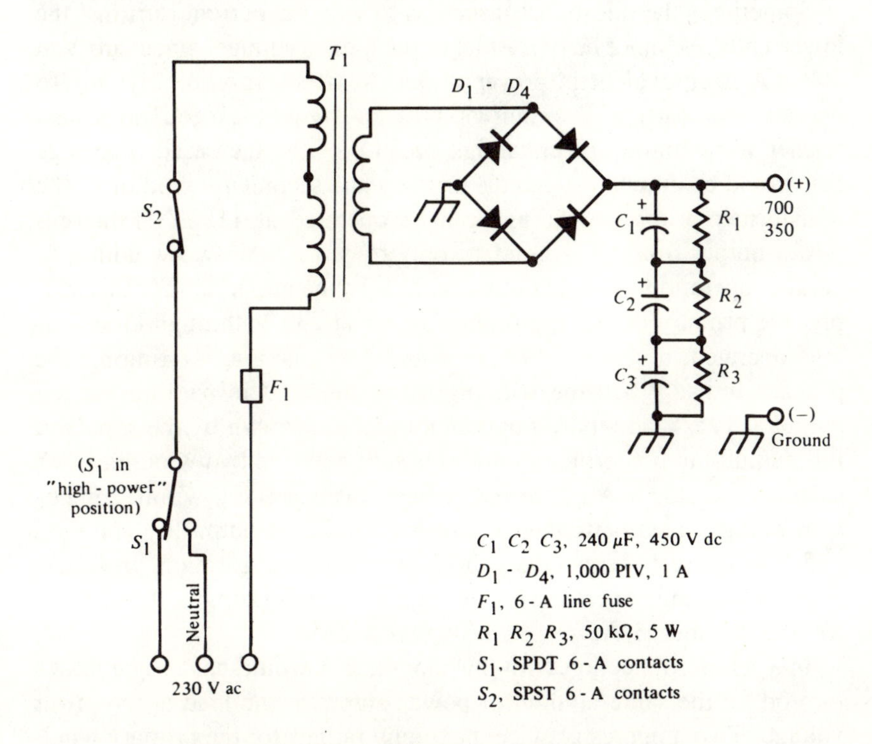

FIGURE 10-5. Power supply with half power switch.

transformer, switching out the 230-V hot lead, and 115 V is now applied to the 230-V windings. With half the normal input voltage at the primary windings, only half the normal voltage appears at the output of the transformer secondary windings. The dc output voltage is reduced by half. This can only be accomplished with a transformer having a primary input requirement of 230 V when standard house current is being used.

A power supply of this type is convenient for use in transmitting circuits for control of transmitter output in a short period of time. By dropping the anode voltage on the final transmitting tube or tubes, power output is instantly dropped. It can be raised back to a normal value again by simply throwing the switch in the other direction. This method of power reduction is much less expensive than adding another transformer or rectifier–filter circuit to an existing power supply to provide a simultaneous and separate half-voltage source. This latter method would require switching of higher voltages on the dc side of the power-supply secondary and possible arcing problems.

Sometimes the question is asked as to why the current rating of the lower voltage cannot be twice that of the higher voltage, since 700 V at 300 mA is equivalent in power to 350 V at 600 mA, or 210 W. To operate this particular circuit at 600 mA would exceed the transformer's maximum power ratings because of the increased amperage that would be drawn through the primary and secondary windings. The transformer primary windings can be operated at 115 V for a full-power output from the secondary only by wiring the two windings in parallel operation. At a dc output of 210 W, the maximum for this supply, the primary draws approximately 1A at 230 V through the windings operated in series. This 1A would be a maximum rating for the primary turns. When the windings are connected in parallel for full output at 115-V ac input, a maximum of 2A is drawn by the supply at full output; but this primary current is split between two windings with each one passing 1 A of current. Power levels are still within component ratings. This particular circuit lowers the dc output by applying 115 V to the series-connected 230-V winding configuration. In series, the two windings operate as one, and the current rating is the same as for 230-V input, 1 A.

In a series connection with this circuit, 2 A would have to be drawn to produce the same amount of power output to the load at the lower voltage. Two amperes is twice maximum ratings for the primary windings connected in series. The transformer is operating high above its maximum ratings and would probably be damaged if operation were continued. Power in watts is not the true rating applied to transformer windings. Current through these windings is the main consideration, and 2-A current looks the same to the windings at 115, 230, 2,000 V, or any value of potential.

This same type of thinking applies to the transformer secondary, which is rated at 300 mA regardless of the voltage produced by the secondary. Here, again, power consumed by the load is not as important a factor as the load resistance or the current demand from the supply.

A 275-VOLT DC-TO-DC POWER SUPPLY

This circuit uses much of the solid-state switching techniques that were discussed in Chapter 8. Low-powered electronic equipment can readily be used in a portable operation using this supply, which derives its input voltage from a 12-V automobile storage battery. The circuit shown in Figure 10-6 is suited to compact construction techniques and delivers a steady output of 275 V at current demands of up to 100 mA.

This transistorized switching circuit utilizes voltage feedback to start the oscillation process and switches the current at a very rapid rate, allowing the use of a small toroidal transformer, which is commercially manufactured by Triad under the designation of part number TY-28. This extremely small, encapsulated transformer has six separate windings that are terminated in numbered contacts around the perimeter of the case. The switching transistors are a relatively common variety and are readily obtained from many different manufacturers under their cross-referenced designations. Although different types of transistors could be used, it is highly recommended that the type specified in the schematic or a proper cross-referenced replacement be used in building this circuit. Other transistors may produce dc voltage outputs that could vary from the design voltage by a factor of as high as 90 percent in either direction.

When the power obtained from a 12-V storage battery is applied to the primary center tap of the power transformer T_1, one of the two transistors will conduct current while the other completely blocks its flow. The voltage feedback at contact numbers 8–11 starts to level off as the conducting transistor reaches its saturation point. When this occurs, the other transistor begins to conduct and the remaining one shuts down. This same half-cycle process occurs again with the second transistor and the cycle is complete. The process starts again and continues

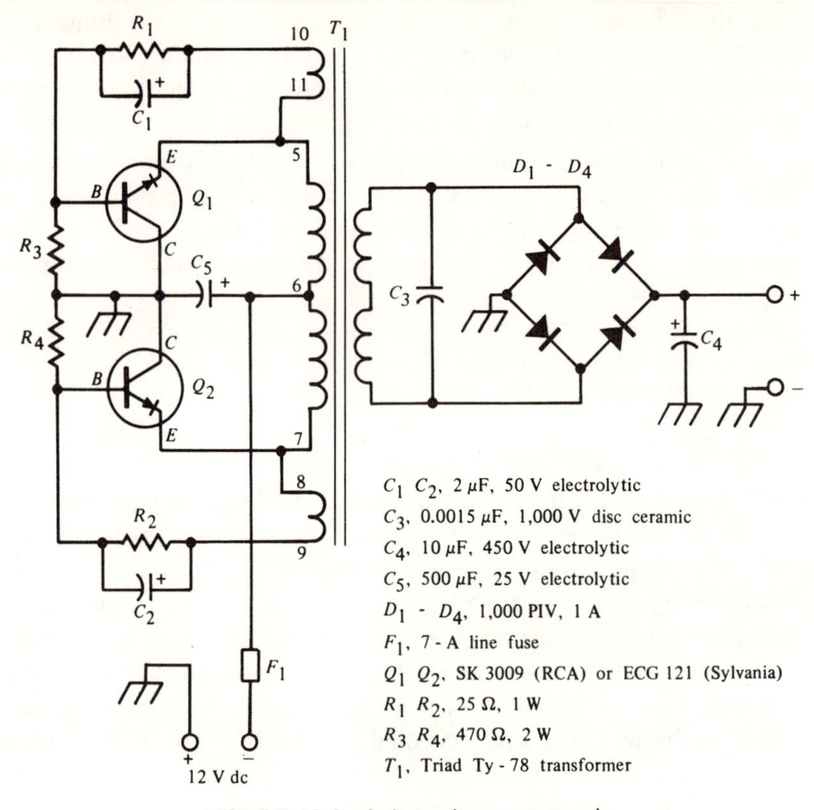

FIGURE 10-6. A dc-to-dc power supply.

until the input voltage source is removed. This is the oscillation process for switching at the transformer primary. The transformer is composed of two windings connected in series that deliver a dc output of approximately 275 V under load when rectified and filtered. Capacitor C_3 is a spike suppressor to prevent possible damage to the solid-state devices within the circuit. The ac output from the unfiltered secondary is a standard square waveform.

Proper operating parameters are established at the base junctions of the two switching transistors by resistors R_1 and R_2. These resistors, which are in parallel with capacitors C_1 and C_2, are waveform shapers and prevent excessive spikes from occurring. Capacitor C_5 stabilizes the input voltage during periods of voltage drop within the external electrical system.

At a dc output of 275 V and 100 mA, the input will draw approximately 3½ A from the storage battery. A 5-A fuse is used to provide protection to the circuit.

The switching transistors used in this circuit must be mounted on a heat sink. Small plastic insulators are often provided with power transistors of this type and are formed to fit the physical shape of these devices. The transistors in this circuit are operated with their cases at ground potential and do not use these insulators, which provide electrical isolation while still conducting the heat from the transistors to the heat sink. Commercially manufactured heat sinks made for these large-cased transistors with four or six cooling fins should be adequate to dissipate the heat buildup within the transistors. Do not attempt to operate this circuit, even in the test stages, without proper heat protection for the transistors. Even operation for a brief period of time can cause permanent damage to the delicate solid-state devices.

Make certain that the source wiring to the power supply is adequate to handle about 4 A of battery current. Conductors that are too small will cause a voltage drop at the transformer center tap under heavy loading conditions, causing voltage drops at the dc output and improper operating conditions on the switching transistors. If any question exists as to the operating voltage at the primary center-tap connection, a measurement with an accurate voltmeter should be taken directly between this point and the circuit ground. The voltage should be a nominal 12 V from an automobile storage battery when no charging is taking place. When the automobile engine is operating, the voltage will read higher. Thirteen and a half volts is the nominal value for a voltage feed from the electrical system while the engine is running. Voltage values from the output of this power supply will naturally be a bit higher when the engine is operating because of the increased primary input.

Because of differences in electrical systems found in automobiles and owing to the slight differences in solid-state components from different manufacturers, if this type of operation is intended, voltage measurements should be taken at the dc output of this supply before operation is attempted with electronic equipment.

Repeating some facts about dc-to-dc power supplies, never use this unit to power high-current devices that require high initial amounts of power to obtain proper operating conditions. Never exceed the maximum current rating of this supply, even for brief periods of time.

If this power supply is used in automobiles for any length of time, dampness and other weather conditions can cause a corrosive buildup on the transistor cases and in other parts of the circuit. A thorough cleaning of components on a regular basis will keep this supply in peak operating condition for a much longer period of time. Solder joints and other connections and contacts should be carefully inspected for breakdown. A corroded solder joint can create a high-resistance point in the circuit and, if located in the primary voltage lead, can heat and further deteriorate, causing large voltage drops and possible failure of operation.

Even though power supplies of this type do not create an excess of noise and interference to audio equipment, a faint tone or whine may be heard in equipment operating from this circuit owing to the oscillation of the switching transistors. This whine may also be heard emanating from the supply itself. This is a normal operating inconvenience and may be disregarded as a potential circuit problem.

A 350-VOLT LOW-CURRENT SUPPLY

A low-cost power transformer rated at 125 V at the secondary winding can be used to build a dc power supply with an output of 350 V under the light loading conditions that might be imposed by a grid cir-

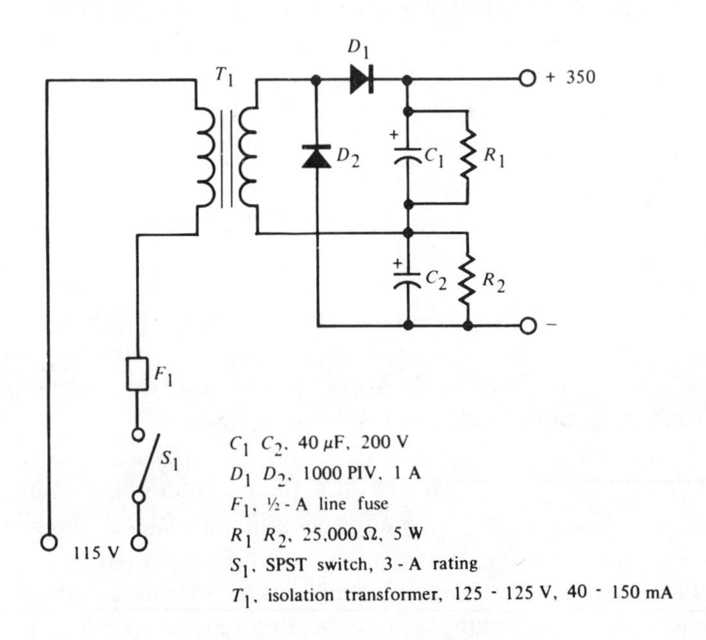

FIGURE 10-7. Low-current supply.

cuit or other low-drain source. Transformers of this type are usually rated from 40 to 150 mA of current drain and can be chosen to meet the expected load requirements.

The size of the power supply depicted in Figure 10-7 makes it very suitable for incorporation into the electronic load so that no external power supply is necessary. The current rating of the power transformer will determine just how small a space the supply will require, as the other components will occupy the same space regardless of the current drain anticipated.

A full-wave voltage-doubler circuit is used to raise the voltage value from the secondary output and was chosen because of its small physical size and decreased cost when compared to a transformer with a higher voltage at the secondary or a full-wave bridge rectifier circuit, which uses twice the number of solid-state rectifiers. The capacitors are rated at 200 V each and are wired in series to present an overall voltage rating of 400 working volts dc. Capacitance value for each capacitor is 40 μF and drops to 20 μF when the two components are combined in series. Two 20,000-Ω, 5-W resistors are used as bleeders across each capacitor. This is one of the simplest circuits that can be constructed to do an adequate job of supplying a voltage in the medium range for purposes of powering low-current loads.

This circuit could be duplicated without using the power transformer by connecting the rectifier and filter circuit directly to the 115-V power line, but the power transformer provides isolation from the line and prevents other problems that are associated with a direct ac line type of power supply.

Primary power is actuated by the control switch S_1. A ½-A fuse protects the circuit. The secondary output feeds the full-wave rectifier circuit, which, in combination with C_1 and C_2, provides an output equal to 2.8 times the secondary ac value.

The value of capacitance in C_1 and C_2 may be increased where space permits, with the higher capacitance values providing a higher degree of dynamic regulation. If the electronic load requires a stiffer control of voltage value, the consideration of higher capacitance may be necessary. Make certain that whatever value of capacitance is used it is equal in each of the capacitors. Capacitors of the same manufacture and type should be used to provide an even division of voltage across each of these components. The bleeder resistors also serve as balancing devices for the capacitors. This resistance in parallel with the capacitor aids the matching process by decreasing any electrical differences within C_1 and C_2.

Finally, this type of power supply is designed to be used with electronic loads that demand only a small amount of current. A transformer that is rated to deliver 100 mA at the secondary voltage of 125 V will deliver only about one third of this amount at the 350-V output from the voltage-doubler circuit, since the dc voltage is almost three times the secondary ac value. Heavy current demands will cause the output voltage to drop considerably and possibly operate the transformer in a region above its maximum ratings.

This circuit may be used with other transformers to provide higher power output. In many instances, it will not be necessary to change any other components in the circuit if the output of these modified supplies is below 330 mA. Higher current drains than this will necessitate changing the rectifiers to units rated at higher than 1 A, which is the maximum current that can be drawn from the ones specified. All other components, however, may remain the same.

This power supply is suited to printed circuit design. The rectifiers, capacitors, and bleeder resistors may all be mounted on a relatively small printed circuit board for an even greater compactness. Owing to the simplicity of a full-wave doubler circuit, the printed circuit board may be either designed and etched by the builder, or a general-purpose board may be purchased from an electronic outlet store and mofidied to make the proper connections to complete this circuit.

A 900-VOLT HIGH-CURRENT SUPPLY

With many types of electronic equipment, there is a need for medium voltage with high-current output. Many tube circuits require this value of voltage and are used in relatively high power output applications. Many of these circuits draw up to 1 A of current, but often this amount is drawn for only a fraction of a second with a longer period of time during which much lower current drains are required. This cooling-off period in between current pulses allows the power-supply components to dissipate the heat that was built up during the high drain period, and actual printed ratings may not be exceeded in these types of applications, because the *average* power drain will be within the ratings of the power supply.

The power supply shown in Figure 10-8 is of heavy-duty construction, with a power transformer rated to deliver 500–600 mA of current to the load at 900-V dc output. This same supply will deliver up to 1 A

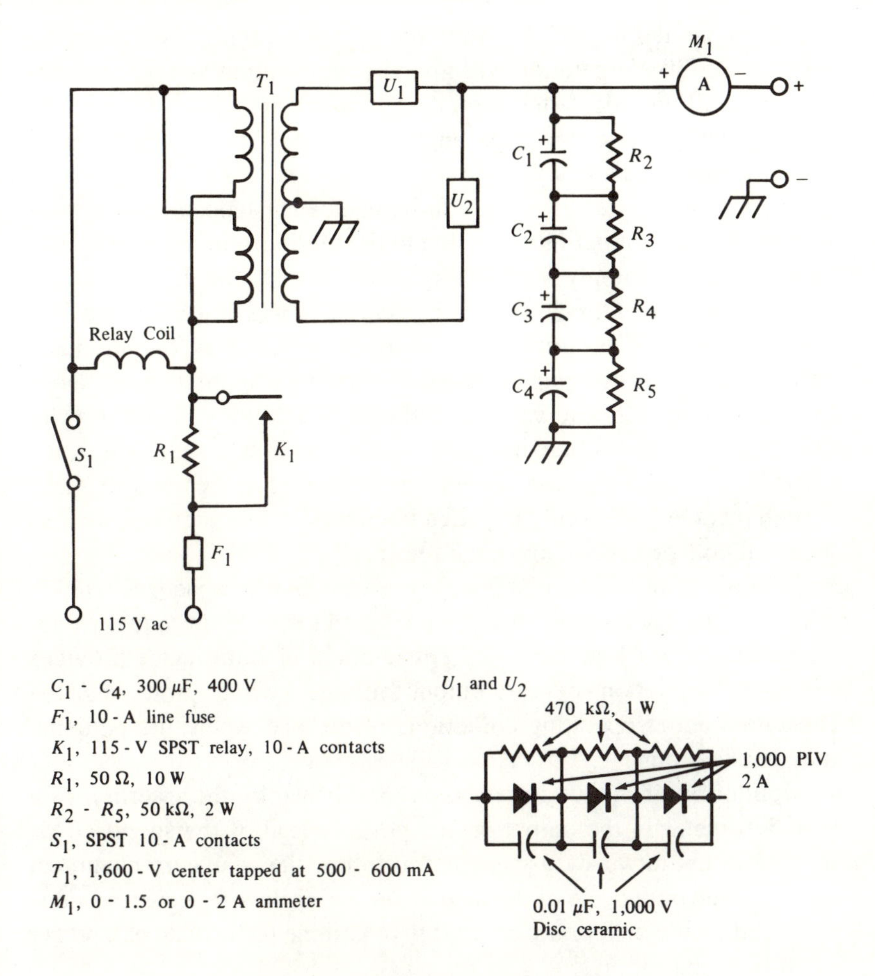

FIGURE 10-8. High-current supply.

to the load during current peaks and still remain within its operating boundaries and offer good voltage regulation.

Alternating-current is fed to the primary through a protective relay circuit that allows partial charging of the filter capacitors before full voltage potential is actually applied to the circuit. A 10-A fuse is used in the primary line to offer overload protection. A 10-A circuit breaker may also be used for this purpose.

Most transformers in the voltage range needed for this circuit have two primary windings that can be connected in series for 230-V operation or in parallel for 115 V. Although a 115-V circuit is shown in the schematic, 230-V operation will give the same output voltage and current and will provide better regulation in many applications where the primary ac lines are not large enough to handle the 9 A of current drawn without causing a voltage drop.

The rectifier circuit is a basic full-wave center-tapped arrangement with a capacitive-input filter. Owing to the high current drains that may be placed on the supply, rectifiers rated at 1,000 V and 2 A of forward current are used. One-ampere-rated diodes would not allow enough safety margin if the circuit were operated at full peak output. Each string of diodes is made up of three components that are wired in series to produce a rectifier equivalent to ratings of 3,000 PRV at 2 A. Protective capacitors and equalizing resistors are wired in parallel with each rectifier to protect the components from voltage spikes and to make certain that they are evenly matched in internal resistance. This assures an equal voltage drop across each rectifier.

The capacitor bank is composed of four 300-μF capacitors rated at 400 V each. The combined capacitance and ratings are equal to approximately 75 μF at 1,600 V. This amount of capacitance provides excellent regulation of the output voltage, while protecting the capacitors under operating conditions of no load where the potential may rise to around 1,200 V.

Monitoring of output current is accomplished by the insertion of a 0–1.5 A meter in the output on the positive lead of the supply. This meter will not move swiftly enough to follow the peak current surges, but when the duty cycle of the load is known, the average current value indicated on the meter may be used to determine the actual peak power drain from the supply.

Four bleeder resistors are wired in series with each other to drain off stored current in the capacitors and are simultaneously wired in parallel with each capacitor to equalize internal differences to make certain that an equal amount of voltage is dropped across each component in the bank. Adequate insulation should be provided around the metal cases of each capacitor. Should the case of the top capacitor come in contact with the circuit ground, a potential of almost 1,000 V will be present while the capacitor is rated at only 400-V insulation. A flashover or arc may occur, destroying the capacitor and possibly damaging the power supply. The bottom capacitor in the bank does not need to be insulated between case and ground, but should be insulated from the next higher capacitor.

The 900-V high-current supply is basically a very simple circuit and should provide trouble-free operation if proper construction practices are used and the power supply ratings heeded. An enclosed cover should be used with this supply to prevent accidental contact. At a voltage potential of this magnitude, this power supply can easily be a lethal device to careless or inexperienced persons.

MULTIVOLTAGE POWER SUPPLY WITH REGULATION

Often there is a need to supply regulated voltages of medium values to an electronic load. While solid-state devices may be used to provide regulation at these voltages, many components are usually required in series connections, because solid-state components are generally low-voltage devices. A much more practical component for voltage regulation at values in the medium range and where very low current drains from these regulated supplies is required by the load is the voltage-regulation (VR) tube. This tube is filled with a gas and presents a constant voltage drop at a determined value just as its solid-state counterpart, the zener diode, does. Voltage-regulation tubes require no external voltage for operation and offer many of the conveniences and advantages of zener diodes, but they are just as susceptible to damage as all electronic tube devices. They are also much larger in physical size than zener diodes.

Voltage-regulation tubes are available for regulated voltages in set values of 75, 90, 105, and 150 V. They can be operated in a series connection to provide higher values. For example, two 150-V VR tubes may be connected in series to provide a 300-V regulator, or a 105- and 90-V series combination may be used to provide a regulator at a value of 195 V.

Voltage-regulation tubes perform their functions by dropping voltage values over their set regulating point through a dropping resistor placed in the circuit in series with the load and in back of the VR tubes, which are shunted to circuit ground. A constant current is required for these devices to operate properly or, at least, a current value that changes by no more than 30 mA from minimum to maximum value. Current drain from most VR tubes should be at least 8 mA and no more than about 40 mA. These gaseous regulator tubes emit a greenish-blue glow when voltage is applied owing to the gaseous interior, which acts as an insulator up to a specific voltage point. When this point is ex-

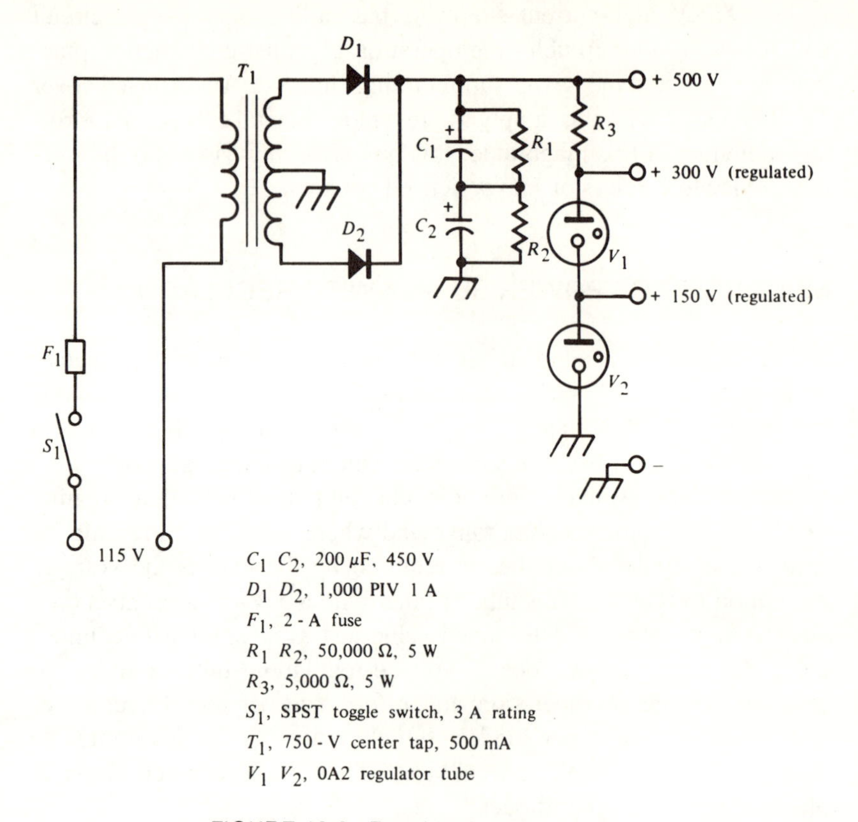

FIGURE 10-9. Regulated power supply.

ceeded, the excess voltage causes current to be conducted to ground. This drain causes a voltage drop across the series resistor, which lowers the voltage to a point just short of the value that will cause conduction within the VR tube. The series resistor is chosen to deliver a normal voltage at the regulator(s) of slightly more than the specified regulation voltage of the tubes. Since VR tubes perform their function by dropping voltage through a resistor, the unregulated input to the tubes must always be more than the regulator voltage drop value.

Figure 10-9 shows a power supply that delivers three separate voltages, 500 V unregulated and 300 and 150 V regulated. This power supply is very straightforward in design, using a full-wave center-tapped rectifier circuit and capacitive-input filter to deliver the 500 V, unregulated. This power source also feeds the regulator circuit, which is made up of a 5,000-Ω, 5-W resistor and VR tubes, each rated at 150

V, in a series connection. The 300-V feed is at the top of VR_1; 150-V is taken from the bottom of VR_1, which, being at the center of the two 150-V regulators, supplies half the series rating of 300 V. All the current being drawn from the three sources originates at the secondary of the power transformer. The total current drawn from the three sources must be within the current limitations imposed on this transformer.

This power supply can provide all the voltages for many types of electronic equipment, such as radio receivers, oscilloscopes, and so on, if a transformer is chosen that has another winding to supply the desired filament voltage. It can be seen that many of the structural disadvantages of tubes are outweighed by the price differential when compared to series-connected zener diodes when using gaseous regulator tubes in medium-voltage circuits.

The actual mounting of VR tubes in the power-supply circuit involves more labor than is required for zener diodes. Tube sockets must be installed, and this usually involves cutting circular holes in the supply chassis. A metal-cutting die is normally used for this purpose. The socket slips into its mounting hole and is then secured by bolts. The correct pins are wired into the power-supply circuit and the tubes inserted.

Power supplies that use VR tubes are more subject to damage from sudden jars to the circuit. The glass envelope of the tube can easily break and cause complete failure of the regulating portion of the supply. Voltage-regulation tubes are, generally, very reliable while operating and rarely if ever fail in an operating condition. Since no filaments are involved, the usual ailment of vacuum tubes, that of an open or defective filament, is not present, and VR tubes will often last the life of the circuit.

QUESTIONS

1. When using a power transformer in a circuit that supplies two dc voltages from the same winding, what considerations must be given to this transformer regarding current drain?
2. A power transformer contains two 115-V windings at the primary. What connection of these windings is used to operate the transformer at a primary voltage of 115 V? at 230 V?
3. In a power supply that delivers both a positive and a negative vol-

tage output, these polarities are either positive or negative in relationship to what?

4. A transformer that has a 125-V primary input and a 125-V secondary output has a turns ratio of what when comparing the primary to the secondary?
5. Draw a schematic of a possible dual-voltage rectifier circuit for use with a center-tapped transformer where one voltage is equal to about one half of the other.
6. What total output power may be drawn from a power supply that has two windings in the secondary, one rated at 400 V at 330 mA and the other rated at 250 V at 100 mA?
7. What is a VR tube?
8. How does a VR tube differ from a zener diode?
9. List the advantages and disadvantages of VR tubes versus zener diodes.
10. When a dc-to-dc power supply is used in an automobile or in adverse weather for lengthy periods of time, what precautions should be taken regarding servicing?
11. How may a filament transformer with a 115-V primary winding and a 12-V secondary winding be used with another power transformer to deliver an output of approximately 130 V using the proper rectifier circuit? Why is this possible?
12. Why do many power transformers have a low-voltage winding of about 12, 6, or 5 V ac?
13. Why do many medium-voltage power supplies designed to power tube-type circuits have a negative voltage output in addition to a positive dc voltage?
14. How may a dc output voltage be varied from a low value to maximum output at the input of a transformer that has two 115-V primary windings?
15. A transformer secondary voltage of 150 V ac will deliver what value of voltage output from a dc supply using a tripler rectifier circuit under conditions of no load?

chapter eleven

HIGH-VOLTAGE POWER SUPPLIES

High-voltage power supply circuits are normally used to supply anode potentials for vacuum-tube circuits operating at high power ratings. Typical operating parameters for transmitting tubes require voltages of 1,500, 2,500, 3,000, and 4,000 V dc and more. These voltage values are usually noncritical, and a high degree of regulation is not necessary. This is true when compared to the regulation that is often required for low-voltage and some medium-voltage circuits. Most of the regulation in these high-voltage circuits will be determined by the size of the filter capacitor or capacitor bank used with the power-supply circuit.

The circuits used for explanation in this chapter may be modified to a high degree simply by changing the power transformer and adjusting the component ratings accordingly. These high-voltage power supplies are probably the simplest circuits discussed so far, but it must be remembered that most of these supplies are lethal to human beings when accidental contact is made. Adequate safety precautions must be taken with each of these circuits regarding proper shielding and chassis isolation. Interlocks should be placed at each entry to the cabinet, as well as other safety arrangements as explained earlier. Don't take any chances with these circuits whatsoever regarding proper bleeder resistor size. Don't depend on the resistance of the applied load to bleed off stored energy in the capacitors. Failure to comply with these circuit requirements will result not only in improper operating conditions to the electronic load but in possible human injury as well.

A 2,000-VOLT, 500-MILLIAMPERE POWER SUPPLY

Figure 11-1 is the schematic of a heavy-duty dc power supply that is designed to deliver approximately 1,000 W to an electronic load requiring a voltage potential of approximately 2,000 V and a current of up to ½ A. This circuit uses a choke-input filter with the inductor lo-

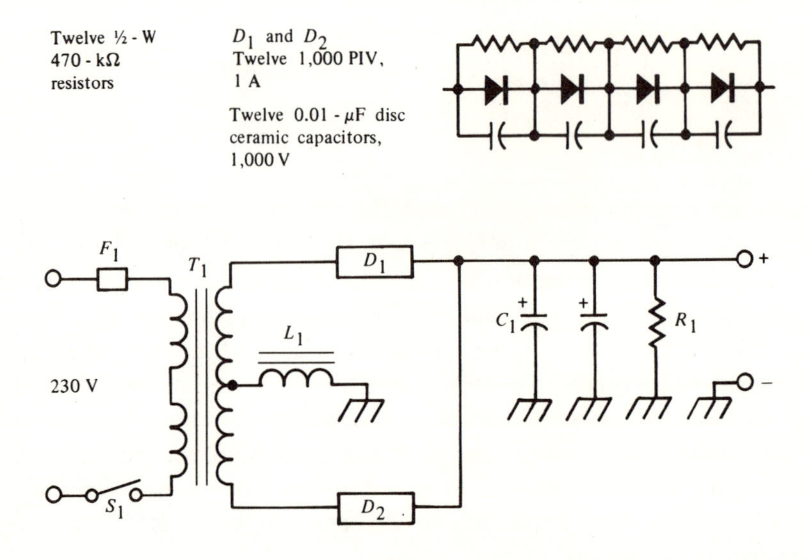

C_1, two 4 - μF, 3,000 - V oil - filled capacitors
F_1, 10 - A line fuse
L_1, 10 - A filter choke, 600 mA
R_1, 100,000 Ω, 100 W
S_1, SPST toggle switch with 20 A contacts
T_1, 4,800 - 5,000 V center tapped

FIGURE 11-1. A 2,000-volt supply using a choke-input filter.

cated in the center tap or negative lead of the power supply for greater reliability. The voltage rating of the choke can be almost any amount above 500 V because it does not see the full-voltage potential, as would a choke that was mounted in the positive leg of the supply.

The power transformer should be rated at approximately 4,800–5,000 V ac across the secondary and have a secondary current rating of at least 500 mA in order to deliver the power output specified. These voltages are available in commercially manufactured transformers, but these tend to be rather expensive when purchased new. Many war-surplus outlets carry transformers in ratings similar to those specified for a fraction of the new commercial prices. Some of these war-surplus units may carry higher current ratings, which can be used to good advantage if more power is needed at the output of the power supply. A 230-V primary line is used in the schematic shown, but 115-V lines may be used if so desired, although regulation may suffer a bit because of the current drawn (almost 10 A at 1,000 W). The transformer shown in the schematic has two primary windings that are connected in parallel for 115-V operation and in series for 230-V operation.

A standard full-wave center-tapped rectifier configuration is used with this supply. Each leg of the rectifier assembly consists of twelve 1,000-PIV diodes rated at 1 A each. The total rating of this assembly is equal to 12,000 PIV at 1 A. The high PIV rating is necessary to prevent rectifier damage, as the PIV measured at the rectifiers is equal to 2.8 times the transformer ac voltage. Equalizing resistors and suppressor capacitors are used in parallel with each diode to prevent voltage overload and damage from voltage transients and spikes.

The filter circuit consists of choke L_1 and filter capacitor C_1. C_1 is made of two 4-μF, 3,000-V capacitors of the oil-filled variety in parallel for a combined rating of 8 μF at 3,000 V dc. Choke L_1 is connected in the negative lead of the dc power supply between the transformer center tap and chassis ground.

The circuit is completed by adding the bleeder resistor, which is rated at 100,000 Ω of resistance and 100 W of power dissipation. This resistor will become very hot in normal operation and should be mounted in a position that will allow a large amount of natural ventilation to cross its surface. Remember that this resistor is wired across the main output voltage line and sees the entire potential of 2,000 V across its contacts. Some type of ventilated cage may be considered to cover this component to allow for natural air circulation while preventing human contact.

Metering of this circuit is easily accomplished by connecting a voltmeter rated to read at least 2,500 V dc between the positive dc output lead and the circuit ground connection. Output current may be read by placing an ammeter rated at 800 mA or more in series with this positive lead, or the meter connections may be reversed and the unit inserted between the transformer center tap and choke L_1. Either of these methods will provide reliable meter readings under conditions of varying load.

This circuit is excellent for uses that do not require a great deal of current variations in normal operation. The output voltage of this dc power supply possesses good regulation, smooth filtering, and a very low starting surge, which would not be the case if a capacitive-input filtering system were used. The output voltage from a supply of this type is usually about nine tenths of the transformer secondary voltage from the center tap to one side, or about 2,160 V using a 4,800-V center-tapped transformer. When the maximum output of 500 mA is drawn, the voltage will drop to a point that is very near the design voltage of 2,000 V. This is a very sturdy supply and is designed to give many hours of continuous service while delivering the maximum amount of power specified.

Transformers of higher current ratings may be used with this circuit to deliver up to 900 mA of output current while still using the same power-supply components. The maximum rating of the rectifier stack is 1 A, which will be well within safe operating limits if operated at 900-mA output.

A 2,800-VOLT POWER SUPPLY

The power supply depicted in Figure 11-2 will supply 2,800–3,000 V at a continuous output current of approximately 400 mA. A capacitive-input filter is used, so instantaneous peak current values of twice the average maximum amount may be drawn. The continuous power rating of the transformer specified in this schematic is about 1,000 W, but under peak current demands the power supply may safely deliver in excess of 2,000 W for brief periods of time.

A full-wave voltage-doubler circuit is used in this power supply. Five 1,000-PIV silicon rectifiers are used in each string, along with parallel-connected capacitors and resistors for equalization and spike-limiting purposes to form each leg, which is rated at 5,000 PIV and 1

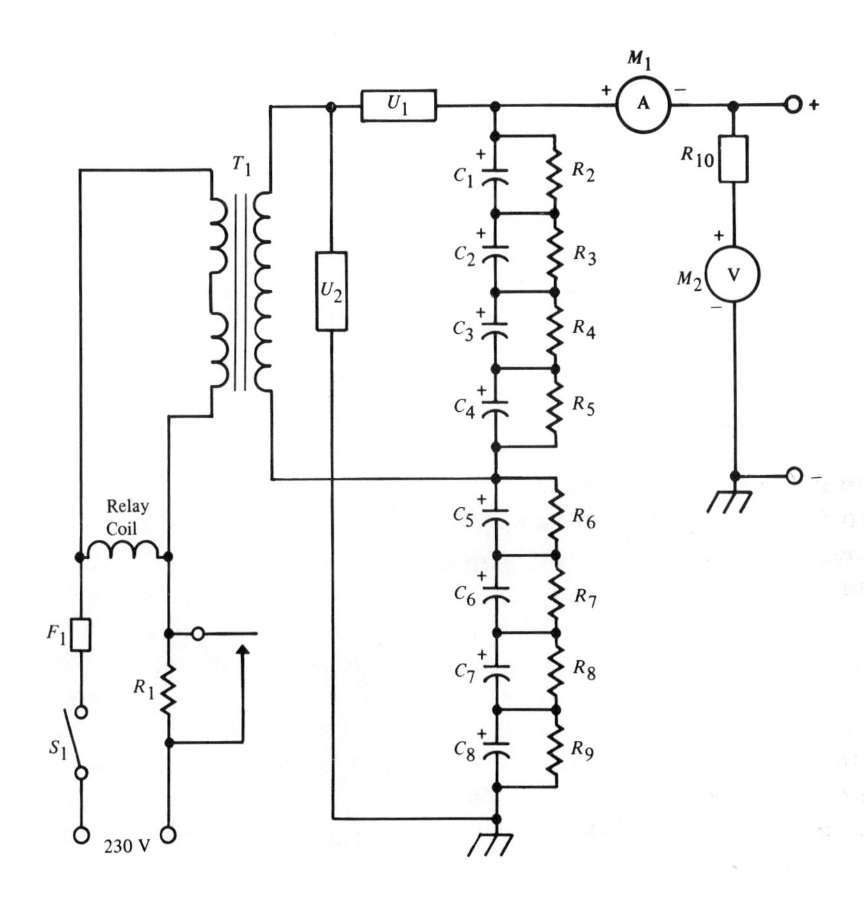

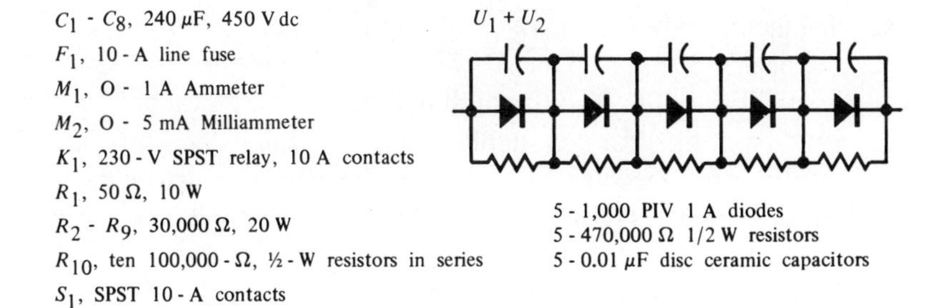

FIGURE 11-2. A 2,800-volt power supply.

A. The capacitive input filter had a very high total capacitance at a total rating of 3,600 V. This capacitor bank is made by combining eight 450-V, 240-μF electrolytic capacitors in a series configuration. In a series connection, capacitors add voltage ratings and divide capacitance ratings. The 3,600-V capacitor formed by this type of series connection exhibits a total capacitance of approximately 30 μF. This is an extremely high amount of capacitance for a power supply of this voltage output, and the dynamic regulation of the supply is excellent under conditions of varying current demand from the electronic load. Equalization resistors are used in a parallel connection across each of the eight capacitors in the bank to assure an equal voltage drop across each component and also to act as a bleeder resistor for each unit. No other bleeder resistor is needed or used in this circuit.

A 230-V primary input ac voltage is recommended for power-supply circuits that supply high amounts of power output. Most of the transformers used for these purposes, however, may be used with 230 or 115 V. The transformer used with this schematic is no exception. The schematic shows a series or 230-V connection in this example.

The secondary alternating current is rectified through the full-wave voltage doubler for an output approximately equal to 2.8 times the ac value under conditions of no load. The bleeder resistors across each capacitor further stabilize the output dc voltage by providing a minimum drain on the power supply. Under conditions of normal to heavy loading, this power supply will deliver between 2,800 and 3,000 V dc to the electronic load. This type of power supply is especially designed to operate well under conditions of highly varying load demands. Voltage drop will not be excessive even when the load current rises instantaneously from a value of approximately 80 mA to a high value of nearly 800 mA. One of the main reasons for this excellent voltage stability is the 30-ΩF combined capacitance available at the capacitive-input filter for this supply. Under conditions of a very heavy load, heating may become excessive in the power transformer, and some arrangements may be necessary to aid the natural cooling provided by normal air circulation. Often, small blowers and chassis fans are used to increase the direct flow of air across the body of the transformer to more effectively remove heat from the core and windings.

Metering is accomplished with this high-powered supply by using an inexpensive 0–5 mA milliammeter in connection with a 1-MΩ resistor used as a meter multiplier. This combination provides an effective 0–5,000 V voltmeter, which keeps a constant monitor on output voltage from the power supply. Another meter, a 0–1 A ammeter is used to monitor the current drain. Power supplies that produce voltage levels and current and power outputs on the order of the one under discussion often fall under legal requirements which state that the voltage and power output must be accurately measured at all times. The meter multiplier resistor must be comprised of very accurate resistors or sizable errors may result at the meter indicator. A string of ten 100,000 μ resistors are used with 5 percent or better tolerances. These ½-W resistors must be specially chosen for accuracy and tested with the meter and compared to another accurate voltage source for final adjustment. Do not use a single 1-M Ω resistor instead of ten 100,000-Ω units. One resistor will break down under the high voltage, and arcing to the meter contacts could occur with a subsequent loss of the resistor and, probably, the meter itself. The multiplier resistor chain is best mounted by wiring each resistor to a printed circuit board or to a punched piece of vector board. This type of material provides good high-voltage insulation and can be easily mounted to the power supply chassis by using ceramic stand-off insulators.

The input of this power supply is conventionally protected from voltage surges and other surge conditions during initial activation by using a charging-delay relay in the input to the primary windings. Adequate interlock and other safety devices should be used to protect all persons who are to operate this power supply.

A 1,500-VOLT HIGH-CURRENT SUPPLY

A power transformer identical to the one used in the previous power supply is used with a full-wave bridge rectifier configuration to produce a power supply with a voltage output of approximately 1,400–1,500 V under a load of about 800 mA continuously and about 1½A under instantaneous peak current conditions (Figure 11-3).

The primary ac input circuit is identical to the previous power supply, but after the secondary winding, the similarity ends. A full-wave bridge rectifier circuit is used to produce a dc output voltage of 1.4 times the rms secondary voltage, or approximately 1,540 V under no-load conditions. This value drops to 1,420–1,500 V under conditions of maximum loading.

The bridge rectifier is composed of diodes that are connected in a series–parallel configuration. Each of the four assemblies is made up

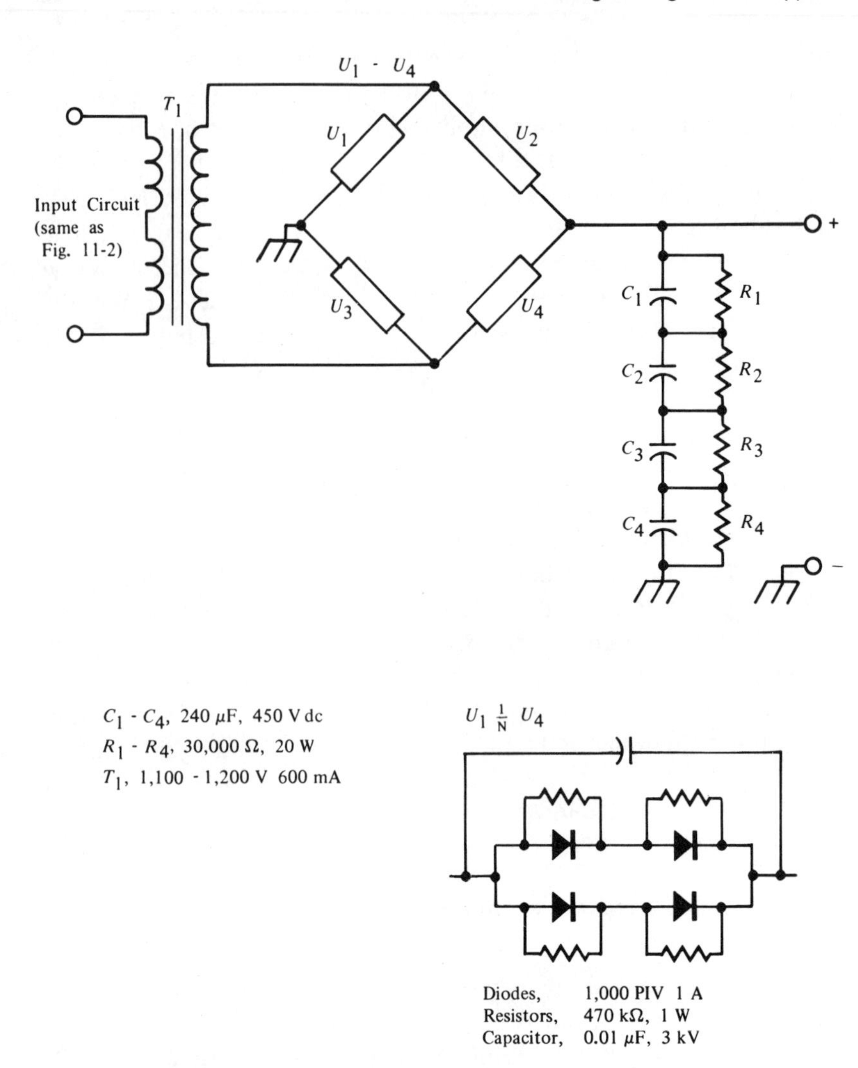

FIGURE 11-3. A 1,500-volt, high-current supply.

of four rectifiers. Two sets of parallel-connected diodes are then connected in series for a combined PRV rating of 2,000 V and a current capacity of 2 A. Actually, it is not necessary to parallel diodes if 2-A units are used. These types of rectifiers are almost as easily found as are their 1-A counterparts, but this circuit, as depicted, shows a good example of parallel–series rectifier circuits. Equalizing resistors are used to assure equal voltage drops across each segment of diodes.

Diodes should be carefully chosen and matched to make certain that the parallel units pass equal amounts of current. For high-current applications, series equalizing resistors are not practical. Each diode may be measured for internal forward resistance with a high-quality ohmmeter or vacuum-tube voltmeter. Forward resistance should be nearly equal for each of the diodes used for these connections.

A capacitive-input filter is also used with this high-voltage power supply and it, too, is very similar to the filter used in the previous power supply, which uses many of the same components. This filter is actually one half of the previous one, as only half the voltage rating is necessary because of the lessened dc output of this supply. Only four of the electrolytic capacitors are used in a series configuration. The total voltage rating of the single capacitor formed by this chain is equal to 1,800 V dc, one half of the rating of the previous filter capacitor bank. However, since only half the capacitors are used as with the previous bank, the total capacitance is doubled. The total combined capacitance of the bank used in this circuit is 60 μF. A filter capacitor of this high value will exhibit excellent regulation characteristics. The dc voltage value will be very stable, even under conditions of tremendous variations in the current demanded by the electronic load.

No provisions have been made for external metering of this power supply, but, again, the previous schematic gives good indications of possible means of monitoring the operating output parameters. A 2-A ammeter may be substituted and placed in series with the positive voltage lead. The same metering circuit may be used to measure voltage as was used with the previous supply, but the readings will always be in the low portion of the meter scale, a situation that is undesirable when accurate readings are necessary. A better metering circuit would use a 0–2 mA milliammeter and the same voltage-multiplier resistor. This will bring the normal indications of voltage into the top quarter of the meter scale.

Many types of electronic loads will require voltage potentials on the order of the values presented by this power supply. Many tubes have a maximum plate voltage rating of less than 2,000 V and generate high amounts of output power by drawing large amounts of current from these power supplies, which deliver a relatively low value of high voltage. Other types of electronic loads produce the same amounts of power-supply circuity best suited to provide the correct values of of 3,000, 4,000, or more volts), and draw lesser amounts of current from the power supply. The specific components used in the electronic load and their ratings in the circuit will determine the type or types of power-supply circuitry best suited to supply the correct values of operating potentials.

In the supply under discussion, a different type of power transformer may be used with the rest of the specified circuitry to deliver slightly different outputs. An 800-V secondary will produce a final dc output of approximately 1,100 V dc under loading conditions. A 1,200-V secondary will deliver approximately 1,600 V at the dc output. A transformer with a tapped primary winding that allows some adjustment of the secondary ac voltage is ideal for this type of power supply and will allow for considerable adjustment of the dc output under many varying conditions of current demand. Alternatively, a variac or other means of varying the input voltage to the transformer may be used to change the dc output. When making substitutions in power transformers, it should be remembered that the filter capacitor chain is rated at a maximum dc value of 1,800 V. If a power transformer has a secondary winding that exceeds the rating of the capacitors, damage could result to these components and possibly the rectifier chain in the circuit specified.

The bleeder resistor in composed of four resistors connected in series with relationship to each other and in parallel with each of the four filter capacitors. This, it will be noted, is equal to one half of the bleeder resistance used in the former power-supply schematic. The power rating of each of the resistors is twice that of the resistors used formerly, and the total combined power dissipation rating of the four resistors is 100 W, or equal to the power rating of the resistors used in the previous supply.

This power-supply circuit was built and designed directly from the power supply mentioned earlier. The power output from both power supplies is equivalent to about 1,000 W continuous and approximately 2,000 W on an instantaneous or intermittent basis. This latter power supply may be used for high-power electronic circuits that require a lower value of voltage; 1,500 V is a standard value for many types of electronic tubes.

The same power transformer may be used to deliver a myriad of other output voltages at power levels of 1,000 and 2,000 W. The rectifier configuration, in most instances, will be the determining factor regarding output voltage and current availability.

Although the filter and bleeder resistor components were changed to make this circuit more economical, all the circuit of the previous supply, with the exception of the ammeter in series with the positive dc output line, could have been used following the full-wave bridge rectifier circuit to obtain the same results of about 1,500 V of output with a current rating of up to 1½ A. The interchangeability of all the high-voltage power supplies of this design offers many advantages.

A 5,000-VOLT POWER SUPPLY

The power supply pictured in Figure 11-4 produces a dc output of approximately 5,000 V under current demands of almost 500 mA. Using Ohm's law, a power output of 2,500 W is possible with this power supply. The output voltage value is extremely high, and this type of supply is used for very specialized applications, such as very high output RF circuits and others.

The circuit is very basic, using a full-wave voltage-doubler rectifier circuit coupled with a power transformer with a 1,800-V ac secondary. Eight diodes are used in each leg of the rectifier chain to produce a rating of 8,000 PIV at 1 A of direct current. This is an adequate rating

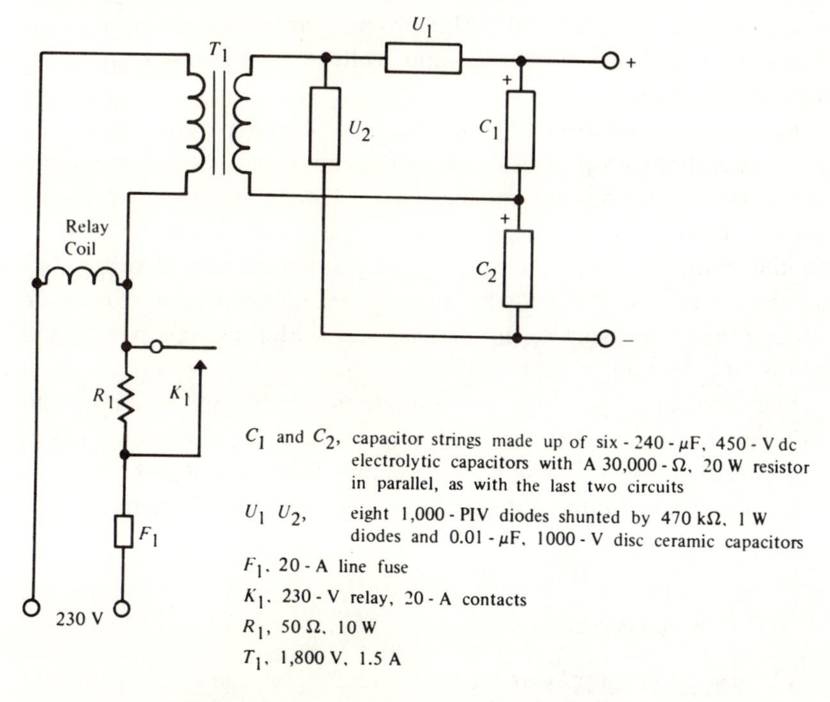

FIGURE 11-4. A 5,000-volt power supply.

to handle the output voltage and current while still maintaining a reasonable safety margin for the components. The full-wave voltage-doubler circuit is chosen for many high-voltage circuits because it is inexpensive and because it adapts readily to standard levels of high-voltage output while using power transformers with common secondary voltages.

The power supply in Figure 11-4 uses a relay protection circuit in the primary leads that delivers a 230-V ac potential to the power transformer. The series resistor in the line substantially drops the amount of voltage that is delivered to the primary windings and, thus, to the filter capacitor after rectification has taken place at the secondary winding. A split second after the circuit has activated, the relay contacts act to short out the series resistor. The circuit then operates normally, but the split-second delay has allowed the filter capacitors to develop a small amount of charge, and a great current surge is avoided, which can cause failure of the solid-state rectifiers.

The circuit pictured uses twelve 240-μF, 450-V electrolytic capacitors to form the two filter capacitors in the doubler circuit. The total combined capacitance value is equal to 20 μF at a voltage rating of approximately 5,400 V dc. This provides an adequate safety factor should a higher than normal ac input voltage be present at the transformer primary windings.

Each capacitor carries its own equalizing bleeder resistor to assure an equal voltage drop across each of the capacitors. Do not substitute one large bleeder resistor for these 12 resistors. Equalizing resistors must be used or the voltage drop across these components may be irregular, with one capacitor having a very small amount of voltage across its contacts while another may drop several times its maximum dc voltage rating, causing an arc or flashover within the component and failure of the supply.

Standard safety precautions are required for this power supply and all others that operate at a voltage potential that is considered dangerous. Adequate spacing should be maintained between all voltage points and the chassis. Voltage of this high value can arc if spacing is not adequate.

HIGHER VOLTAGES

The last power supply under discussion operates at a voltage potential higher than that used for most electronic applications other than in the commercial field where voltages many times higher may be required. But higher voltages are possible using the same schematics presented in this chapter by simply substituting other power transformers and increasing the component ratings specified to a value that will be adequate to handle the new operating parameters presented. As dc voltages pass the 10,000-V value, problems can occur on the human safety level regarding magnetic fields presented by these high values and X-ray radiation, which can also present a hazard when working in areas where extremely high voltages are present. These very high voltages fall in a realm that lies outside of the scope of this text. Although the principles for generating these voltages are very similar to the theory already discussed, the problems that are associated with these high voltages require special study and attention.

QUESTIONS

1. What danger exists in a power supply that has recently been activated even after all primary power is removed? What should be done to protect the technician from this condition?
2. What can occur in a high-voltage power supply if the filter capacitor(s) does not have a voltage rating higher than the applied voltage?
3. Why can the rectifiers of a power supply be damaged by current overload during initial power-supply activation if adequate precautions are not taken at the transformer primary input?
4. What is a secondary or backup bleeder resistor? What is its purpose in the overall circuit?
5. Why are insulation properties important in high-voltage power supplies?
6. What factors determine the type or types of choke arrangements that will be used at the output of the rectifiers?
7. What factors should be considered in the physical mounting of the main bleeder resistor in a high-voltage power supply?
8. What are the expense advantages of using a full-wave doubler circuit for high-voltage rectification?
9. What are the advantages of using a variac at the primary input to a high-voltage power-supply transformer?
10. Name the two types of power-supply safety interlocks. Explain how each is used.

appendix a

SCHEMATIC SYMBOLS

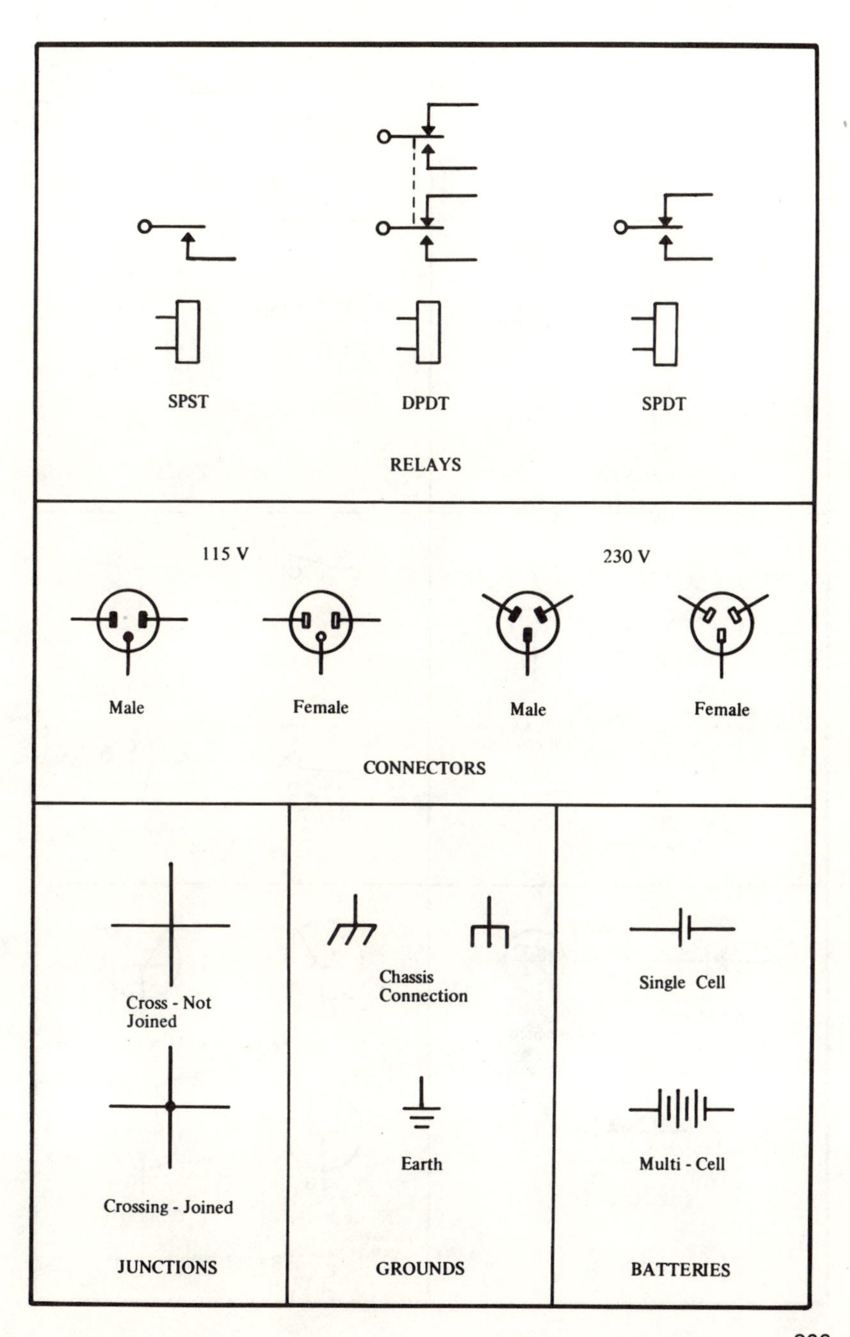

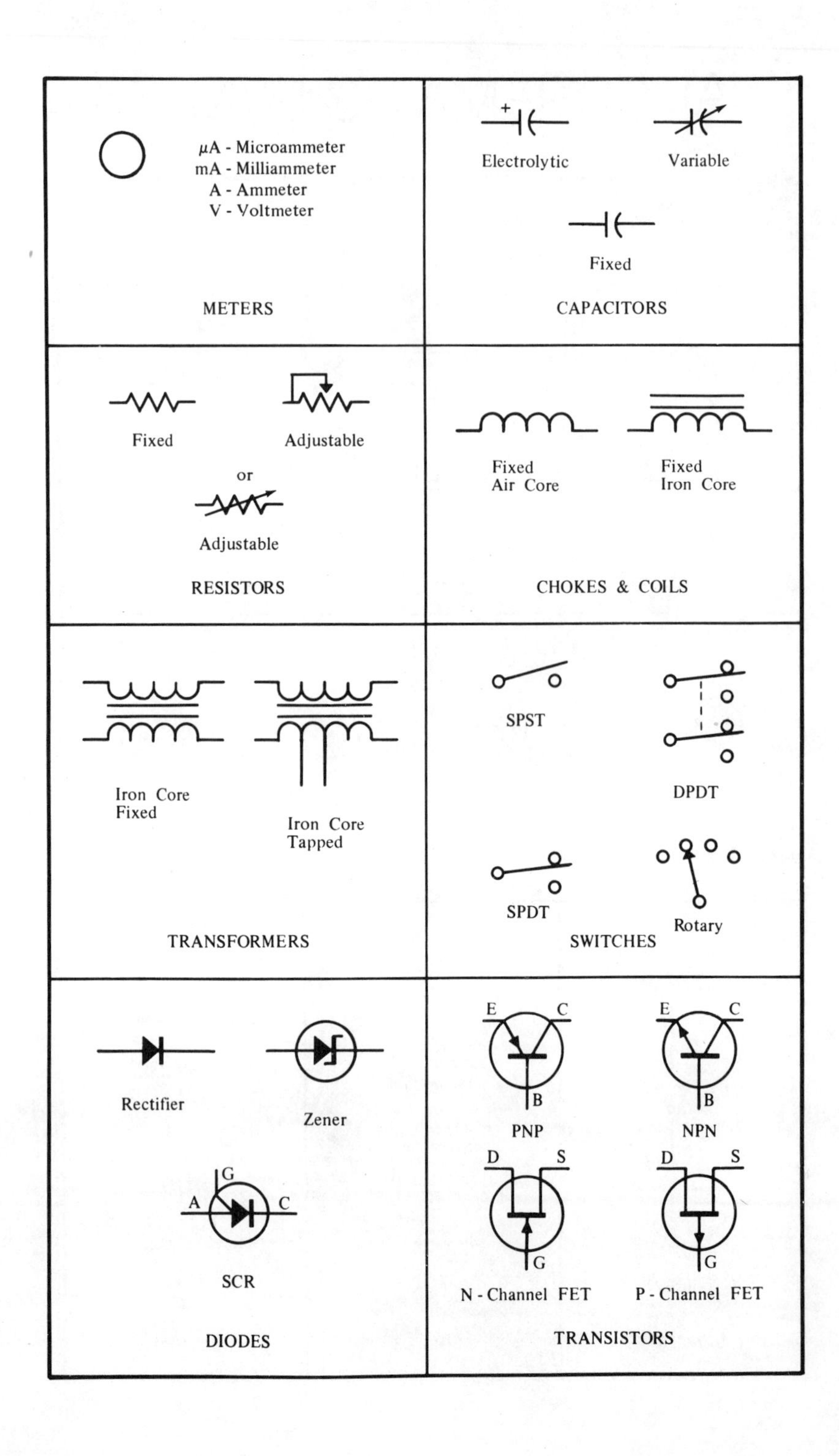
μA - Microammeter
mA - Milliammeter
A - Ammeter
V - Voltmeter
METERS
+
Electrolytic
Variable
Fixed
CAPACITORS
Fixed
Adjustable
or
Adjustable
RESISTORS
Fixed
Air Core
Fixed
Iron Core
CHOKES & COILS
Iron Core
Fixed
Iron Core
Tapped
TRANSFORMERS
SPST
DPDT
SPDT
Rotary
SWITCHES
Rectifier
Zener
G
A
C
SCR
DIODES
E
C
B
PNP
E
C
B
NPN
D
S
G
N - Channel FET
D
S
G
P - Channel FET
TRANSISTORS

appendix b

STANDARD METAL GAUGES CHART

Gauge No.	*American*	*U.S. Standard*[2]	*Birmingham*
1	0.2893	0.28125	0.300
2	0.2576	0.265625	0.284
3	0.2294	0.25	0.259
4	0.2043	0.234375	0.238
5	0.1819	0.21875	0.220
6	0.1620	0.203125	0.203
7	0.1443	0.1875	0.180
8	0.1285	0.171875	0.165
9	0.1144	0.15625	0.148
10	0.1019	0.140625	0.134
11	0.09074	0.125	0.120
12	0.08081	0.109375	0.109
13	0.07196	0.09375	0.095
14	0.06408	0.078125	0.083
15	0.05707	0.0703125	0.072
16	0.05082	0.0625	0.065
17	0.04526	0.05625	0.058
18	0.04030	0.05	0.049
19	0.03589	0.04375	0.042
20	0.03196	0.0375	0.035
21	0.02846	0.034375	0.032
22	0.02535	0.03125	0.028
23	0.02257	0.028125	0.025
24	0.02010	0.025	0.022
25	0.01790	0.021875	0.020
26	0.01594	0.01875	0.018
27	0.01420	0.0171875	0.016
28	0.01264	0.015625	0.014
29	0.01126	0.0140625	0.013
30	0.01003	0.0125	0.012
31	0.008928	0.0109375	0.010
32	0.007950	0.01015625	0.009
33	0.007080	0.009375	0.008
34	.0.006350	0.00859375	0.007
35	0.005615	0.0078125	0.005
36	0.005000	0.00703125	0.004
37	0.004453	0.006640626	
38	0.003965	0.00625	
39	0.003531		
40	0.003145		

appendix c

WIRE SIZE, CURRENT RESISTANCE CHART

Wire Size A.W.G.	Diameter in Mils	Circular Mil Area	Feet per Pound, Bare	Ohms per 1,000 ft 25°C	Current-Carrying Capacity at 700 C.M. per Amp	Diameter, mm
1	289.3	83,690	3.947	0.1264	119.6	7.348
2	257.6	66,370	4.977	0.1593	94.8	6.544
3	229.4	52,640	6.276	0.2009	75.2	5.827
4	204.3	41,740	7.914	0.2533	59.6	5.189
5	181.9	33,100	9.980	0.3195	47.3	4.621
6	162.0	26,250	12.58	0.4028	37.5	4.115
7	144.3	20,820	15.87	0.5080	29.7	3.665
8	128.5	16,510	20.01	0.6405	23.6	3.264
9	114.4	13.090	25.23	0.8077	18.7	2.906
10	101.9	10,380	31.82	1.018	14.8	2.588
11	90.7	8,234	40.12	1.284	11.8	2.305
12	80.8	6,530	50.59	1.619	9.33	2.053
13	72.0	5,178	63.80	2.042	7.40	1.828
14	64.1	4,107	80.44	2.575	5.87	1.628
15	57.1	3,257	101.4	3.247	4.65	1.450
16	50.8	2,583	127.9	4.094	3.69	1.291
17	45.3	2,048	161.3	5.163	2.93	1.150
18	40.3	1,624	203.4	6.510	2.32	1.024
19	35.9	1,288	256.5	8.210	1.84	0.912
20	32.0	1,022	323.4	10.35	1.46	0.812
21	28.5	810.1	407.8	13.05	1.16	0.723
22	25.3	642	514.2	16.46	0.918	0.644
23	22.6	510	648.4	20.76	0.728	0.573
24	20.1	404	817.7	26.17	0.577	0.511
25	17.9	320	1,031	33.00	0.458	0.455
26	15.9	254	1,300	41.62	0.363	0.405
27	14.2	202	1,639	52.48	0.288	0.361
28	12.6	160	2,067	66.17	0.228	0.321
29	11.3	127	2,607	83.44	0.181	0.286
30	10.0	101	3,287	105.2	0.144	0.255
31	8.9	80	4,145	132.7	0.114	0.227
32	8.0	63	5,227	167.3	0.090	0.202
33	7.1	50	6,591	211.0	0.072	0.180
34	6.3	40	8,310	266.0	0.057	0.160
35	5.6	32	10,480	335	0.045	0.143
36	5.0	25	13,210	423	0.036	0.127
37	4.5	20	16,660	533	0.028	0.113
38	4.0	16	21,010	673	0.022	0.101
39	3.5	12	26,500	848	0.018	0.090
40	3.1	10	33,410	1,070	0.014	0.080

appendix d

RESISTOR COLOR CODING CHART

Color	*Significant Figure*	*Decimal Multiplier*
Black	0	1
Brown	1	10
Red	2	100
Orange	3	1,000
Yellow	4	10,000
Green	5	100,000
Blue	6	1,000,000
Violet	7	10,000,000
Gray	8	100,000,000
White	9	1,000,000,000
Gold	—	0.1
Silver	—	0.01

ANSWERS

CHAPTER ONE

1. (a) Half-wave rectifier: characterized by a single diode acting on one-half of each ac cycle.
   (b) Full-wave, center-tap: two diodes, each acting on one half of the ac cycle for full-wave output. Circuit ground is obtained at the center tap of the power transformer secondary.
   (c) Full-wave, bridge: identified by four diodes in the circuit configuration with ground obtained at the junction of the top and bottom diode assemblies, as shown schematically.
2. A half-wave rectifier conducts during only one half of the ac cycle, either the positive portion or the negative portion. The full-wave rectifier contains two rectifiers or two sets of diodes that conduct during each half of the cycle and combine the direct current from each portion of the cycle.
3. A sine wave.
4. Current that is of only one polarity, plus or minus, which varies from a low value to a high value.

5. Alternating current varies from a low voltage to a high voltage, passes through zero voltage, and then repeats its low–high fluctuation at a reverse polarity. Pulsating direct current varies only from low to high and does not change polarity.
6. Choke input and capacitor input.
7. Capacitive-input filter circuits tend to deliver higher voltage dc outputs with the no-load voltage rising to a very high value. Choke-input filters normally produce a lower dc voltage output and maintain this output over large fluctuations of current demand.
8. While operating into a specified load.
9. It is a mathematical computation derived from the voltage–current ratio.
10. A specific minimum value of inductance for a particular circuit under a specific load.
11. A standard choke has a fixed value of inductance, whereas a swinging choke changes its inductance value as circuit demands are changed to maintain the correct value of inductance.
12. To serve as protective devices by draining or bleeding off stored current in filter capacitors, and to maintain a minimum load for the power supply.
13. Stored current in the filter capacitors may remain at a potentially lethal state for long periods of time, presenting this high potential at the dc output terminals, which may come in contact with a person.
14. Average voltage is about 115–230 V, frequency is 60 Hz and single phase.
15. Less current is drawn from the higher voltage line to produce the same power output. Electric conductors are subject to less heating effects and voltage drops.
16. With a multitapped transformer primary winding, a constant voltage transformer, or a variable voltage transformer.
17. Transformer, rectifier, filter.
18. A rating applied to a specific component, which is the difference between its recommended operating parameters and its absolute breakdown parameters.

19. The primary winding is the line voltage input portion of the transformer.
20. Transformation, rectification, and filtering.

CHAPTER TWO

1. The power transformer is usually constructed from sections of laminated plates composing a core with two mutual windings, the primary and the secondary.
2. Alternating current.
3. A voltage will be induced in the secondary winding only for the instant it takes for the voltage to reach its peak. When the voltage has peaked, excessive heating will occur in the primary winding.
4. To change the voltage either to a higher or lower value than the ac line voltage. In some isolation applications, the secondary voltage may be the same as the ac input voltage, with the secondary voltage being completely isolated from the primary line source.
5. When applied to power transformers, this is the maximum safe rating for transformer operation on a continuous basis unless indicated otherwise.
6. Voltage and current. The volt–ampere rating is arrived at by multiplying the voltage output times the current drawn from the transformer.
7. CCS, continuous commercial service; ICAS, intermittent commercial and amateur service. The former indicates continuous operation; the latter indicates intermittent uses.
8. A rectifier is a device that will pass current in only one direction.
9. Silicon and germanium.
10. This is the maximum reverse voltage a rectifier can withstand during the portion of the ac cycle where it does not conduct.
11. The I_{rep} rating is generally used to determine the maximum allowable current to be safely passed by a diode. I_s ratings apply to rectifier ratings upon initial circuit activation where the current may soar to many times the I_{rep} rating for a fraction of a second.
12. They are many times smaller, exhibit less internal resistance, do

not require a separate power supply, and are generally less expensive.

13. A rectifier stack is a combination of several diodes wired in series or parallel for greater current or voltage ratings.
14 Thermal effects are measured in heat that is thrown off by the diode when it passes current. Protection may include metal heat sinks or even small fans and blowers in high-current applications.
15. Transient voltage spikes may occur due to distant lightning strikes or to switching of high-current devices on the same power line. Diodes may be protected by effectively bypassing these spikes with capacitors in parallel with the ac line, or by using Thyrector diode assemblies.
16. Parallel capacitors across each diode, series resistors in the transformer secondary, or a relay-controlled resistor in the ac line to the transformer primary windings.
17. In a series circuit, high-value resistors should be paralleled or shunted across each diode rectifier in the string. In parallel circuits, low-value resistors are placed in series with each diode.
18. They smooth out the ripple component, producing a pure dc output.
19. Electrolytic capacitors have a high value of capacitance for a physically small amount of space. They are usually polar devices with plus and minus terminals.
20. Series: 100 μF and 600 V; parallel: 400 μF and 300 V.
21. Adequate insulation should be provided between the metal cases of the capacitors and ground to prevent high-voltage flashover.
22. The bleeder resistor.
23. To provide a minimum load for the power supply and to drain stored voltage in the filter capacitors to prevent accidental shock.
24. The capacitors in the filter would retain the full potential of direct current for a prolonged period of time, making the supply hazardous even with the ac line disconnected.
25. Choose a resistor with a more than adequate power rating for its application. Small-wattage, high-value resistors should also be wired in parallel with each bleeder resistor to provide a source of current drain should the main bleeder(s) fail.

CHAPTER THREE

1. Aluminum is often too flimsy for very large components without work with while maintaining a high degree of structural strength. Aluminum is often too flimsy for very large cmponents without resorting to extra bracing. Steel is very strong and can usually withstand even the largest components without showing signs of stress. Steel is much heavier and more difficult to work with regarding the cutting of slots and holes and, in general, custom designing to suit special circuits.
2. The amount and size of the components to be used and their total weight.
3. All components should be laid out on a flat surface and arranged in the order in which they will appear on the chassis. When this has been accomplished, the total area covered can be determined by measuring. These dimensions can be applied to the chassis needed to provide a mounting surface for these major components.
4. They should be of the metal-cutting variety, which drills a much neater hole without causing extreme bending or warping of the chassis metal.
5. A metal-cutting bit should be used, and the hole begun at a relatively low drill speed. As the bit begins to stabilize, the speed can be increased. Only moderate pressure should be applied to prevent bending of the chassis. After the hole is drilled, it should be filed smooth to prevent tearing of conductor insulation.
6. Metal-cutting dies or punches in various sizes and shapes.
7. The affected area may be pressed in a large bench vise or pliers may be used to straighten out the severe bends.
8. They should be mounted near one of the corners to take advantage of the better stressed areas.
9. Aluminum stress bars may be used to brace the chassis against a metal bottom plate or against the sides of the chassis.
10. Resin-core varieties.
11. A cold-solder joint is caused by solder being dropped on a joint instead of being allowed to flow into the heated joint elements. A cold-solder joint does not provide a good electrical connection.

12. A good mechanical joint should always be formed before the iron is applied. The conductors or contacts to be soldered should be free of foreign materials. Once a mechanical joint has been formed, the soldering iron or gun is applied to the work surface and not to the solder itself. When the work surface becomes hot enough, solder is applied and allowed to flow over the surface of the conductors or contacts. The finished joint is then allowed to cool without any physical movement and is then examined for the desired shine and rigidity.
13. A good mechanical joint.
14. Some means of dissipating the heat buildup before it reaches the crystalline interior of the devices is necessary. This can be accomplished with a pair of long-nosed pliers held firmly against the component lead between the soldering area and the component case, or by using a special heat sink device that clips on to the lead.
15. Safety to human beings, which is determined by the amount of coverage provided by the cabinet or cover. All the circuit should be completely immune from accidental contact.
16. Vacuum-tube voltmeter, volt–ohmmeter, inductively coupled ac ammeter, transistor checker, capacitance meter.
17. Degree of accuracy.
18. It is a device that measures the alternating current in a power line. The device is inductively coupled to the ac line, and no direct connection is required.

CHAPTER FOUR

1. The point between conductance and nonconductance through the diode.
2. Voltage ranges from a fraction of a volt to several hundred volts, with power ratings of from a fraction of a watt to around 50 W for discrete components.
3. Diodes, zener diodes, transistors, thyristors (SCR'S).
4. To act as a voltage reference for the sensing circuits to act upon.
5. It serves to stabilize the value of the output voltage during periods of varying current demands.

6. A poorly regulated supply can cause voltage-sensitive circuits, such as oscillators, phase-locked loops, and other frequency-determining circuitry, to drift off the design frequency.
7. A series-regulator circuit uses a control transistor in series with the dc output. Regulation is efficient and transistor heat dissipation tends to be high.
8. A shunt regulator uses a control transistor in parallel with the dc output. Dissipation is lower than with a series regulator and efficiency is not as high.
9. A series regulator circuit generally provides a higher degree of voltage regulation.
10. A shunt regulator provides adequate voltage regulation while maintaining a lower cost figure for the components required to complete the circuitry.
11. High-current power supplies usually require transistors and other solid-state devices with the ratings to dissipate a large amount of heat and to pass large amounts of current. Mounting of these components on very large heat sinks is usually required.
12. A variable-voltage regulated power supply has the advantage of being useful for many different circuits that require different values of regulated voltage. For transmitting purposes, a variable-voltage supply can be used to set the output power of the equipment being powered.
13. A current regulator determines the maximum amount of current that may be conducted into the circuits under power.
14. With current regulation, damage to powered circuits is avoided and a steady state of operation is more easily obtained regardless of outside changes such as line voltage and load.
15. A switching regulator provides a high amount of regulation while keeping component dissipation to a low value.
16. The transistor is either in a full-on or full-off condition, which tends to dictate cooler operations.
17. A transistor switching regulator is usually most useful in a dc-to-dc application with low to medium amounts of load current demand. A thyristor-type regulator is best suited to ac-to-dc applications with medium to high amounts of current demand.
18. For very high current applications, thyristors provide the device rating at a lower cost than their transistor counterparts.

19. High-voltage and or high-current applications usually require SCR devices.
20. To suppress the transients or noise that is generated through the on–off switching process.

CHAPTER FIVE

1. $280 \times 1.4 = 392$ V.
2. 630 divided by 1.4 divided by 3 = 150 rms V.
3. 500 mA or 0.5 A.
4. A voltage-doubler circuit will allow the use of a transformer having half the ac rms voltage across the secondary windings. The lower voltage transformer will have less windings in the secondary, usually a smaller physical size, and should cost less.
5. 120 Hz.
6. 60 Hz.
7. Multipliers with even multiples are balanced regarding their half-wave components, resulting in a 120-Hz ripple frequency. Odd-multiple circuits have one leg that is unbalanced, and this half-wave 60-Hz component is reflected in the output.
8. A full-wave doubler is composed of two half-wave rectifier circuits that charge one of two capacitors connected in series across the dc output during alternate half-cycles. The charged capacitors combine their respective voltages in the series circuit for a total dc output of twice the peak ac value.
9. A voltage tripler of the full-wave variety uses one set of two diodes to form a half-wave voltage doubler to charge one of the two output capacitors to twice the peak rms value, and a single half-wave circuit to charge the other capacitor to the peak rms value. The two output capacitors, one at twice the peak voltage and the other at the peak voltage value, combine to produce an output of three times the rms peak value.
10. Half-wave doublers.
11. If the ac ground lead is inadvertently connected to the positive lead or ungrounded lead of the voltage multiplier, any appliances with grounded metal cabinets will carry the full potential of the dc output of the multiplier, which will be present be-

tween the cabinet and ground of the multiplier. This occurs because one side of a half-wave multiplier circuit is always connected directly to the ac line.

12. Only in noncritical home entertainment devices that normally have plastic cabinets to prevent shock. The popular ac–dc table radios are a good example of such devices.
13. 150 divided by 1.4 = approx. 107 rms V; 107 × 1.25 = approx. 134 V dc.
14. Series aiding.
15. 20 μF; 40 × 40 divided by 40 + 40 = 20.

CHAPTER SIX

1. Measurement is often required by law. Measurements indicate whether or not the power supply is operating within safe tolerances. Metering indicates operation, to an extent, of the electronic load.
2. By the passage of current through a moving coil.
3. D'Arsonval.
4. The strength of the field produced by the permanent magnet. The number of turns in the moving coil. The opposing strength of the hairspring.
5. The hairspring opposes the meter movement to a point that dampens the needle in its travel over the scale. This prevents the sudden force created when current passes through the moving coil from driving the indicator to a point that is higher than the actual reading.
6. Horizontal mounting of the meter scale.
7. The ampere.
8. A voltage multiplier.
9. The upper half.
10. It channels a portion of the current around the meter contacts while allowing another portion to travel through the moving coil.
12. In some instances with an ohmmeter, but in most, with an external resistive bridge.
13. Across the output of the power supply.

14. In series with the power supply and the load, either in the positive or negative lead.
15. Half the resistance of the moving coil or internal resistance of the meter. If the internal resistance is 50 Ω, the shunt resistance should be 25 Ω.
16. Meter switching will allow both values to be measured. For current measurements, the meter is placed in series with the load by the proper switch contacts; for voltage measurements, the meter leads are switched across the load.
17. Broken meter face plate. Cracked meter case. Any signs of arcing.
18. Multiplier resistors are often of the high-voltage type or resistor stacks are made up of many small resistors for voltage and arcing protection.
19. Not normally. Because many peaks of current are so fast that the indicator cannot travel that swiftly. By the time the coil gets the instantaneous pulse to move the needle, the pulse has ended its duration and current value is back to a low point again.
20. Light-sensing devices are coupled with a small light source within the meter case. When the needle interrupts this light source by traveling in front of it, blocking the rays from the light-sensitive device, the device conducts or stops conducting, depending on the type, and a small relay is energized or interrupted.
21. The number eight.
22. Light-emitting diode (LED) and liquid quartz. Light-emitting-diode displays produce their own illumination by passing current through solid-state material. This is a cold light, producing no heat. Liquid-quartz displays reflect light over certain portions of their surface area by energizing those areas when current is passed through the display.
23. Digital displays give clear, concise readings. They are easier to read in most applications than conventional meters. Disadvantages include higher initial cost, they require a separate power source, and they are more susceptible to RF interference. They are not suited to measurement of continually varying inputs.
24. Power-supply meters can have high amounts of voltage connected directly to their inputs. Often the needle indicator carries

the full voltage potential. A broken plate could bring dangerous contacts into a position that could be accidentally touched by the operator.

25. A moving coil within a meter is the device that produces the magnetic field as current flows through it. The indicator is normally attached directly to the moving coil. As the coil moves with the current that passes through it, its attached indicator also moves, giving an indication of voltage or current on the meter scale that it travels over.

CHAPTER SEVEN

1. Interlocks, automatic shorting bars, circuit breakers, and fuses.
2. Interlocks are installed in power supplies at access covers and, depending on the type, open or close their contacts when this cover is opened or removed. The primary ac voltage to the supply may contain an interlock that will open contacts when an access cover is removed, thus interrupting the alternating current flow to the transformer primary. Another type of interlock that closes its contacts when the cover is removed may be placed in series with the filter capacitor positive terminals and chassis ground to short circuit any charge the capacitor may have stored.
3. They should be checked with an ohmmeter for proper switching and the contacts should be cleaned or burnished regularly to assure good electrical conduction.
4. An automatic shorting bar may tend to loose its natural spring effect or tension after repeated uses. This may cause the bar to remain in the same position, away from the capacitor terminals, even when the cover is removed, instead of shorting to the capacitor.
5. Because safety devices perform a function that technicians are supposed to do. If it appears that a safety device is properly working in the power-supply chassis, the careless technician may depend on it to do his job for him, causing a serious accident.
6. Damages can occur to almost every component in the power supply. Silicon rectifiers are more subject to this damage, however, than most of the other components. Damage occurs because of the tremendous surge of current that is present when an access

cover is removed from an entrance that is protected by an interlock that shorts the filter capacitor positive terminal to chassis ground. Excess current is drawn through the primary and secondary windings of the power transformer, through the rectifiers, and through any other series component between the primary line and the filter capacitor(s).

7. The primary factors include the type of line and the amount of alternating current that is to be drawn by the operating dc power supply.

8. A direct short in a low-power circuit may cause the drain on the line to be less than the fuse blow-out point. The current drain could be sufficiently damaging to the dc power supply components to cause severe heating and a possible fire, although the fuse remains intact.

9. Fuses or circuit breakers should be installed in the two hot leads only. No fuse or any other device should be placed in the neutral lead. If the neutral is fused, should it blow out, there is still a 230-V potential across the two hot leads.

10. Electrically, a voltage that is of less value than the primary working voltage may be delivered to the transformer, causing low or erratic dc output and improper operation in regulated circuits. Regarding safety, a line cord that is too small to pass the current demanded by the power supply will become hot, severely so in many cases, and can cause a fire.

11. All equipment can be shut down from one location. In case of an accident, all voltage in the accident area can be terminated, stopping any further danger to an accident victim or to persons giving aid.

12. Insulation of these tools must be adequate to withstand the voltage potentials that they are likely to come in contact with.

13. An insulated shaft lessens the possibility of shorting components within the power supply, especially when work in close places is expected. It has an added safety feature for the user in keeping a possible voltage point at the end of the screwdriver when contact is made instead of just below the insulated handle in an uninsulated shaft variety.

14. This technique prevents any electrical current that might be encountered accidentally from traveling through the heart muscle

by a path that extends up one arm, through the chest, and down the other arm.

15. Normally, no. This is true because ac voltage potentials are still present at the switch. Power has been interrupted from the transformer primary, but a hazard of electrical shock from the primary line still exists within the power-supply chassis or cabinet.
16. Mental and physical fatigue, personal problems that detract from the work being performed, and overconfidence.
17. Technician's trance is a state of being that comes over many technicians who have worked long hours on one project that involves detailed work, often around live voltage points, and in poor lighting conditions. This condition can be cured by taking a short 15- or 20-minute break from the continuous work and thinking about other things to relax the mind and body.

CHAPTER EIGHT

1. A circuit that changes dc input to ac output.
2. By switching the polarity of the dc voltage at a rate that will drive a power transformer with a center-tapped primary winding when a vibrator or transistorized switching is used.
3. Vibrator, transistor, silicon-controlled rectifier.
4. A small vibrating reed is pulled in one direction by a magnetic coil. This action causes the reed to come in contact with a stationary reed, which causes current to flow between the two elements. This contact automatically cuts power to the magnetic coil, releasing the vibrating reed, which swings to the other stationary contact, causing current to flow in the opposite direction. The process energizes the magnetic coil again and the cycle repeats itself over and over.
5. A square wave.
6. A center-tapped primary.
7. 60 and 120 Hz. Specialized units are sometimes made that will switch at a rate of 400 Hz.
8. Vibrator circuits depend on the mechanical movement of a reed relay to switch voltage. Moving parts in the vibrator cannot offer the same reliability as the nonmoving switching offered by

solid-state devices. Voltage arcing occurs across the vibrator contacts, causing interference to audio devices and voltage drops within the circuit. This is not present in solid-state circuits. Vibrator circuits offer less efficiency than solid-state circuits.

9. A nonsynchronous vibrator has only two stationary reed contacts for switching current to the power transformer primary. A synchronous vibrator has an extra set of contacts that switch the output voltage of a center-tapped secondary winding on the power transformer, rectifying this voltage for a dc output to the filter circuits.
10. 65 percent maximum; $6 = 15 \times 0.65 = 58.5$ W.
11. High-current devices such as motors that require a high starting current, which may be as much as 20 times the normal operating current. Devices that require a pure sine wave to operate properly will give erratic performances when driven by the square-wave output presented by *most* inverter circuits that are not filtered to deliver true sine-wave output.
12. Better efficiency, less noise interference, higher power output due to their ability to handle higher amounts of input current than most vibrator-driven inverters. Smaller size and lighter weight.
13. As much as 90 percent and more in many circuits.
14. Solid-state switching rates may run to many thousands of cycles. At these switching rates, smaller coils that are constructed of a lightweight material may be used with better efficiency.
15. Powered iron that is sprayed with an insulating lacquer after the crushing process. This insulates each tiny particle from the others. These particles are then pressure assembled into a circular core that resembles a donut.
16. Higher input voltages are more suitable for SCR circuits. Voltages to 750 V may be effectively used with current inputs of around 25 A. Transistor circuits may draw current of over 100 A, but usually operate at input potentials of less than 100 V.
17. A square wave. A sine wave.
18. Current feedback and voltage feedback.
19. Where output power levels, load current, and frequency demands are high.
20. Any of the conventional types that are used with ac-powered supplies.

21. They are very much subject to damage from sudden current surges. If normal operating current drain is 100 mA and a surge occurs that peaks at 500 mA, this five times increase in the secondary creates a five times increase in the primary, which may draw several amperes in normal operation. When this surge causes current to pass through the transistors that is considerably higher than their maximum ratings, destruction of the devices is almost certain.
22. $50 \times 10 = 500$ W. The input supply voltage has nothing to do with the question asked.
23. Transistors draw less current at higher voltages to produce the same amount of power output that would be produced at lower voltage values of input. This results in cooler operation of these solid-state devices and longer, more dependable life of the unit.
24. $20 \times 32 \times 0.90 = 576$ W.
25. 400 V and 800 mA in parallel.
26. $225 \times 1.4 \times 3 = 945$ V. $550/4.2 =$ approx. 130 mA ($1.4 \times 3 = 4.2$).
27. Their operation is inferior to solid-state units in almost every way, including voltage stability and output, current output, frequency stability, and size and weight.

CHAPTER NINE

1. To provide a source of resistance for the conducting zener diode to drop voltage across. When the diode conducts, voltage is dropped through the resistor because of the added current drain. This voltage drop is thrown off as heat from the resistor.
2. It is the vacuum-tube equivalent of the zener diode and regulates voltage in the same manner. VR tubes drop voltage of specific values that are preset in the manufacture of the tube. While most zener diodes have internal voltage drops of relatively low values, VR tubes drop voltages on the order of 100 V and more.
3. The battery.
4. The control transistor or other solid-state element will become overheated in normal operation and will cause erratic operation of the electronic regulation circuit and will probably fail completely in a short period of time.

5. To prevent undesirable hums, noises, and oscillations in transistorized electronic equipment. Cross modulation and erratic operation can also result in many types of solid-state equipment, in addition to frequency drift in electronic circuits.
6. The problem is probably due to a filter capacitor that is of too small a value to provide the proper dynamic regulation of the output voltage. Also, the current drain at peak demand periods may be more than the power-supply components were manufactured to deliver.
7. The unregulated voltage to the electronic regulation circuit may have dropped below the critical point due to a defective transformer, capacitor, or rectifier giving the electronic regulation circuit no excess voltage to drop in the circuit. A component may have failed in the electronic regulation circuit such as a diode, transistor, or possibly a series or parameter-setting resistor.
8. Voltage-regulation tubes, zener diodes, silicon-controlled rectifiers, and integrated circuits.
9. To handle a larger amount of current through the circuit without excessive dissipation in either of the series control elements.
10. This sets a value on the control of the voltage value under various loading conditions. Large voltage fluctuations are said to indicate poor efficiency; while minute changes indicate good efficiency.

CHAPTER TEN

1. That the total power drain from both dc outputs does not exceed the total power output rating of the transformer secondary windings.
2. Parallel; series.
3. In relation to the circuit ground point.
4. One to one.
5.

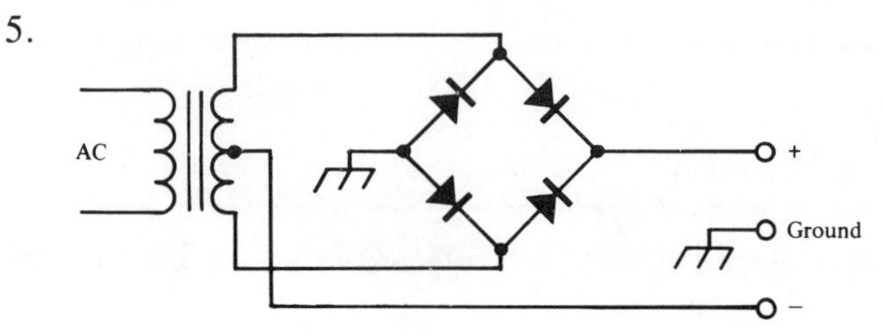

6. Approximately 157 W.
7. A device that presents a constant voltage drop and is used in the regulation of voltages from about 70 to 150 V dc or more when used in series connections.
8. A VR tube performs the same functions as a zener diode but is constructed of materials other than crystal as is found in the zener diode. The VR tube contains a glass shell that has elements for dc voltage connections. A gas is contained in this glass shell and insulates the two connections up to a certain voltage where conduction starts through the gas from the negative connection.
9. Voltage-regulation designed to regulate at much higher voltage, and one VR tube costs less than the many zener diodes that would be required to regulate the same voltage. Disadvantages include constant current drain requirements, low total current drain, and less immunity to accidental jars and vibrations.
10. A corrosive buildup may occur after continued use in adverse conditions, which can cause breakdown of solder joints and improper circuit operation. A thorough inspection should be done regularly along with regular cleaning of circuit contacts.
11. The filament transformer may be used in a configuration that uses the 12-V winding as the primary. A power transformer with a 12-V ac secondary winding is necessary. This latter winding is connected to the 12-V winding of the filament transformer. The output will be 115 V ac. This voltage may then be rectified using a bridge circuit for an output of approximately 130 V dc under light loading conditions. This type of operation is possible because transformers are reversible ac devices.
12. This winding is used to supply ac filament voltage to the tube-type devices within the electronic load. Most tubes require one of these standard voltages.
13. The negative output voltage is usually required by the grids of the tubes in the curcuit.
14. The windings may be fed with a 115-V source. For low voltage output from the supply, the primary windings would be switched to a series connection. A 230-V input would be required in this type of connection to supply the properly rated secondary voltages, but 115 V is being applied and the secondary voltage is about one-half of the normal value. The dc output will also be

equal to one half of the normal value. For maximum output, the primary windings are switched to a parallel circuit where 115 V is required to supply full secondary output. The resultant dc output from the power supply is approximately double the former value.

15. 630 V dc.

CHAPTER ELEVEN

1. Stored electrical energy in the capacitors could still be present at lethal levels if the bleeder circuit has failed. This situation can be eliminated by shorting the capacitor terminals with an insulated screwdriver or shorting bar.
2. The unit can fail due to an electrical arc between the two capacitor plates, which causes a short circuit. This usually damages the capacitor permanently, and can also destroy rectifiers and other components in the circuit.
3. When the power supply is first turned on, the filter capacitor has no charge and forms a dead short in the circuit for a fraction of a second while charging is taking place. As the capacitor charges, the resistance increases and the circuit returns to normal operating conditions. This effect can be visually demonstrated by holding the probes of an ohmmeter across the terminals of a filter capacitor. With the resistance range in the high-ohms position, the needle indicator will immediately go to 0 Ω upon initial contact, and then will begin to rise to a high-resistance level as charging from the ohmmeter power source takes effect.
4. It is a small, high-resistance carbon resistor that is wired in parallel with the main bleeder resistor to act as a source of power-supply load in case the main bleeder fails.
5. At potentials of very high voltage, electrical arcing can occur. In extremely high areas of potential, arcing can occur across air gaps. Insulation properties are especially important at high voltages to prevent damage to components and to allow for safe operator contact.
6. The power-supply work load requirements will determine which type of filter circuit is needed or desired.
7. Insulation and ventilation are main requirements in the physical

mounting of the bleeder resistor(s). The component(s) should be placed in a position where natural ventilation can remove the heat. The bleeder should also be inaccessible to touch to prevent serious electrical shock.

8. Only half the number of rectifier diodes are required over a standard full-wave bridge circuit, and a transformer with half the secondary voltage ratings may be used to create the same output voltage.
9. The variac provides a means of adjusting the primary voltage input to the power transformer, which also has a direct effect on the dc output voltage. Power output from transmitting loads may be directly controlled by controlling the dc voltage output with the variac control.
10. One type is built with normally open contacts; the other has normally closed contacts. Normally open varieties open their contacts when triggered and are usually placed in the primary voltage line to interrupt primary power to the power transformer. Normally closed varieties are placed between the high-voltage output lead and circuit ground. When a door or access is opened, the device shorts the voltage to ground, triggering a circuit breaker or blowing a fuse. Both types render the circuit cold or without voltage when triggered.

INDEX

D

E

F

G

H

I

L

M

N

O